THE HOOFPRINTS GUIDE TO

The Wild Horses *of* Cumberland Island

THE HOOFPRINTS GUIDE TO

The Wild Horses *of* Cumberland Island

Written and Illustrated by

Bonnie U. Gruenberg

QUAGGA PRESS

The Hoofprints Guide to the Wild Horses of Cumberland Island

ISBN 13: 978-1-941700-17-4

Library of Congress Control Number: 2015952706

Published by Quagga Press, an imprint of Synclitic Media, LLC
1646 White Oak Road • Strasburg, PA 17579 • www.quaggapress.com

Also by the author

The Wild Horse Dilemma: Conflicts and Controversies of the Atlantic Coast Herds (Quagga Press, 2015)
The Hoofprints Guide Series (Quagga Press, 2015)
Assateague • Chincoteague • Corolla
Ocracoke • Shackleford Banks • Cumberland Island
Essentials of Prehospital Maternity Care (Prentice Hall, 2005)
Birth Emergency Skills Training: Manual for Out-of-hospital Providers (Birth Guru/Birth Muse, 2008)
Hoofprints in the Sand Wild Horses of the Atlantic Coast (as Bonnie S. Urquhart; Eclipse Press, 2002)
The Midwife's Journal (Birth Guru/Birth Muse, 2009)
Hoofprints in the Sand: Wild Horses of the Atlantic Coast, Kindle Edition (Quagga Press, 2014)
Wild Horses of the Atlantic Coast: An Intimate Portrait, Kindle Edition (Quagga Press, 2014)

Forthcoming

Wild Horse Vacations: Your Guide to the Atlantic Wild Horse Trail with Local Attractions and Amenities (Quagga Press, 2015)
Vol. 1: Assateague, MD | Chincoteague, VA | Corolla, NC
Vol. 2: Ocracoke, NC | Shackleford Banks, NC | Cumberland Island, GA
Wild Horses! A Kids' Guide to the East Coast Herds (Quagga Press, 2015)
Birth Emergency Skills Training, 2nd Edition (Synclitic Press, 2015)

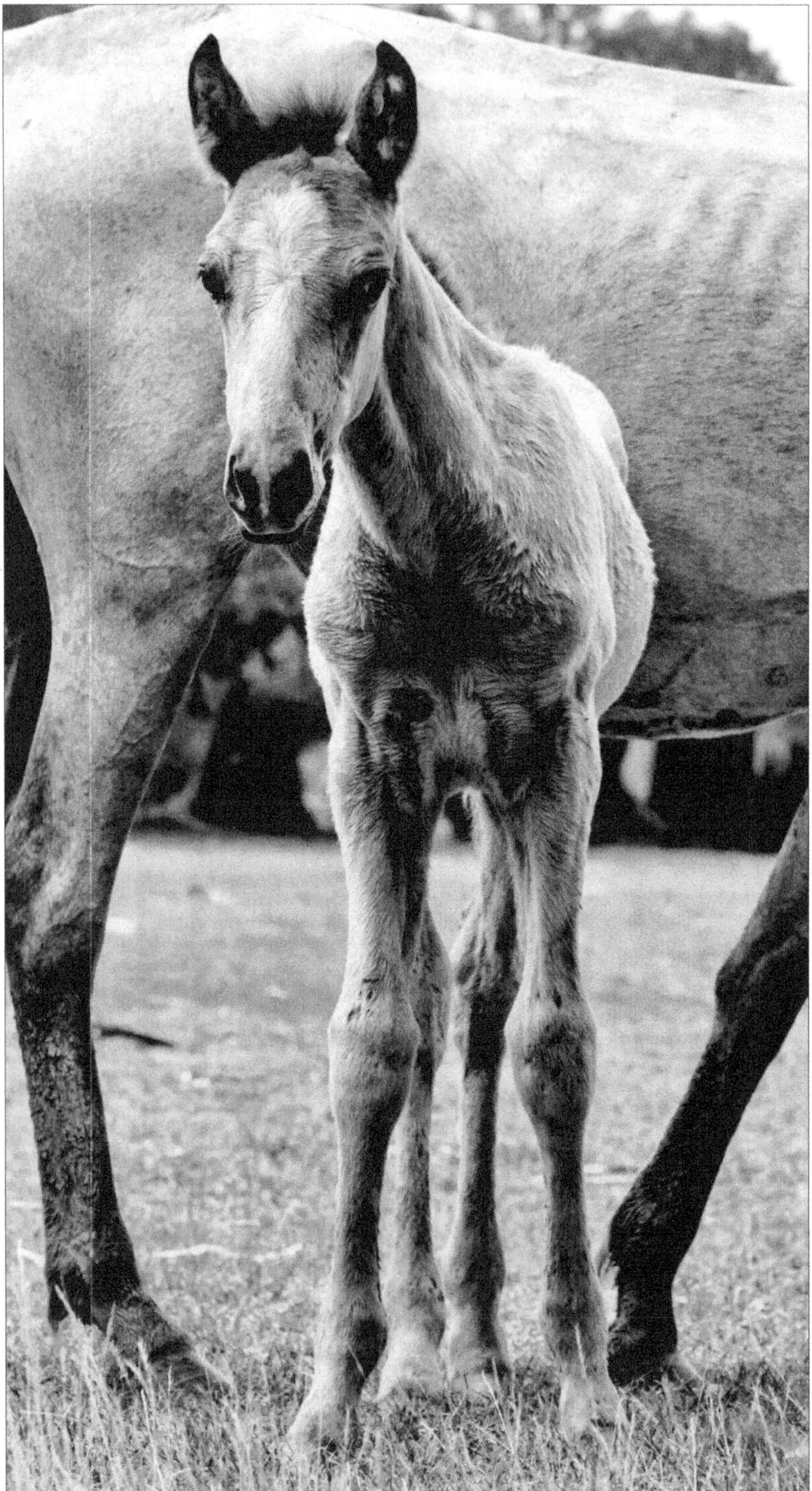

Introduction

National Park Service wildlife biologist Doug Hoffman reads tracks as effortlessly as Indiana Jones reads hieroglyphics. We were negotiating the circuitous Grand Avenue, which was more a broad trail than a road, discussing the pressing environmental concerns of Cumberland Island National Seashore and the influence of free-roaming horses. Periodically he would leave the truck to peer critically at subtle impressions in the moist soil, where horse, hog, deer, armadillo, and bird had left their marks in the forest's guest book. He moved sinuously, with the self-assurance of a man accustomed to living alone in the wilderness.

Doug's duties as a wildlife biologist on Cumberland Island National Seashore keep him busy around the clock and around the calendar. At the time of our meeting, he had been awake more than 26 hours, keeping vigil on the moonlit beach, dispatching feral hogs bent on destroying the nests of sea turtles. From there he had moved on to loggerhead patrol, cruising the beach for "turtle crawls," indicators that new clutches of sea turtle eggs had been deposited in the night. By the time he had collected me for a tour of the equine herds and habitats, a lesser man would have been crippled by sleep deprivation. Somehow, his mind remained quick and his demeanor pleasant.

Unlike the other American coastal islands with free-roaming horses, Cumberland Island is mostly lush, dense subtropical forest of saw palmetto, magnolia, cedar, holly, pine, and myrtle. Majestic 300-year-old live oaks stand like wizened sentries along the pathways, Spanish moss trailing wispily from their limbs. Broad swaths of light filter through the dense forest canopy to illuminate the confusion of saw palmettos below. Many of the horses, however, were thin, and their habitat overgrazed. Political pressures have prevented the Park Service from implementing any sort of management plan to limit their numbers, so nature keeps the balance in its own way. Foals are more likely to die young on Cumberland Island, and in its relentless quest for food, the herd can pressure the ecosystem and place stress on native species.

When I first started researching the wild horses in the mid-1990s, I was surprised to find that wild horses lived on a number of Atlantic barrier islands and had once ranged along much of the Atlantic coast. They made their first hoofprints there not long after the arrival of early European settlers, and in time they ran free on innumerable North American islands and peninsulas from the Caribbean to Canada. I learned that small herds remained on the coast of Virginia, North Carolina, Maryland, and Georgia; on Sable Island, off Nova Scotia, Canada; and on Great Abaco Island in the Bahamas. Each population of horses has its own character, its own history, and its own set of problems. In most cases, these animals have made a unique contribution to local history, and each herd has its own detractors and defenders.

After my first book, Hoofprints in the Sand: Wild Horses of the Atlantic Coast, was published in 2002 by Eclipse Press, I dove in deeper, interviewing experts, evaluating the evidence, and monitoring the herds. I explored management conflicts that encompass political, economic, and cultural issues as well as purely scientific ones. I studied storms and shipwrecks, equine behavior and genetics, history, epidemiology, barrier-island dynamics, sea-level rise, beach development, and the perpetual clash of viewpoints. I pored over hundreds of documents, from historical papers to scholarly articles to court transcripts, so that I might accurately present the pertinent issues. Distilling all this information, I tried to present all sides of the issues fairly so that readers might reach their own conclusions. The result is The Wild Horse Dilemma: Conflicts and Controversies of the Atlantic Coast Herds (Quagga Press, 2015) the most comprehensive work ever published about these horses.

Wild Horse Dilemma is exhaustively researched, copiously documented, and peer-reviewed; but at 600 pages, it may be too long for many people eager to learn about a particular herd. For readers with limited time, I created the Hoofprints Series. Excerpted from Wild Horse Dilemma and containing additional photographs, each Hoofprints book presents a single Atlantic Coast herd in sufficient detail to satisfy both the layman and the academic.

I take all my wild-horse photographs though telephoto lenses that let me keep my distance. When horses approach, I retreat. My goal has been to remain so peripheral to their lives, they will forget that

I am nearby. Because countless people have stroked them, fed them, and lured them, some can be momentarily docile, occasionally indifferent, or routinely bold and pushy in the presence of people. As anyone bitten or trampled can attest, they are no less wild than horses that avoid human contact. When we impose ourselves and our desires on their lives, when we habituate them to our presence, when we teach them to come to us for food and attention, we rob them of their wildness. When we treat them as we would their domestic counterparts, we miss the opportunity to observe them in a natural state, that is, to appreciate the things that make them irresistibly attractive. We miss the very point of driving past thousands of their tame kin to seek them out. We create something like a petting zoo hazardous to us and to them. If we truly love and respect wild creatures, we must learn to stand back and enjoy watching them from afar. Only then can they—and we—know the real meaning of wildness.

As the earth's dominant species, we have the power to preserve or destroy the wildlife of the world and the ecosystems in which they live. The choices we make regarding wild horses are far-reaching. We alter their destiny whether we act or choose to do nothing. We can begin to deal wisely with wild horses by understanding the facts and discovering how the threads of their existence are woven into the tapestry of life. Only through understanding can we hope to make rational, educated decisions about the welfare of these fascinating, inspiring animals.

Bonnie U. Gruenberg
Strasburg, Pennsylvania
October 8, 2015

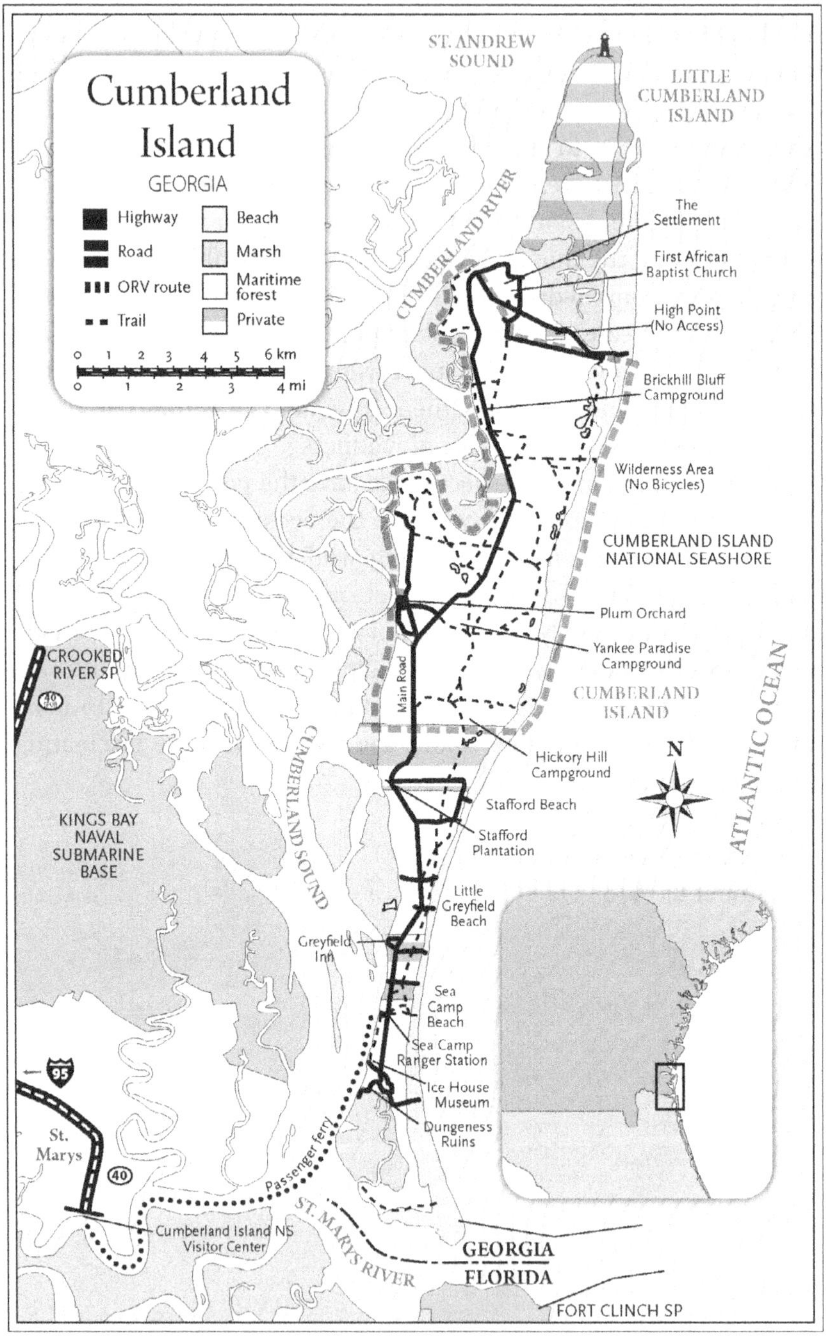

Cumberland Island
GEORGIA
Highway
Road
ORV route
Trail
Beach
Marsh
Maritime forest
Private
0 1 2 3 4 5 6 km
0 1 2 3 4 mi
ST. ANDREW SOUND
LITTLE CUMBERLAND ISLAND
CUMBERLAND RIVER
The Settlement
First African Baptist Church
High Point (No Access)
Brickhill Bluff Campground
Wilderness Area (No Bicycles)
CUMBERLAND ISLAND NATIONAL SEASHORE
Plum Orchard
Yankee Paradise Campground
CUMBERLAND ISLAND
ATLANTIC OCEAN
Main Road
Hickory Hill Campground
N
Stafford Beach
Stafford Plantation
CROOKED RIVER SP
40
KINGS BAY NAVAL SUBMARINE BASE
CUMBERLAND SOUND
Little Greyfield Beach
Greyfield Inn
Sea Camp Beach
Sea Camp Ranger Station
Ice House Museum
Dungeness Ruins
95
St. Marys
40
Passenger ferry
ST. MARYS RIVER
Cumberland Island NS Visitor Center
GEORGIA
FLORIDA
FORT CLINCH SP

At first glance, the bay mare with the crooked blaze appeared to be a free spirit. When all of the other horses in the band were content to graze beneath the moss-laden live oak trees in the heat of the day, the mare was off in search of greener pastures. It was not difficult to see what drove her. She was hungry. Her ribs were prominent, her hip bones jutted, and individual vertebrae were discernible across her topline.

Her young foal, an Appaloosa with a white blanket across his hips, learned from an early age that his mother lived life on her own terms, and if he wished to be part of her activities she required him to keep up. If he wanted to nurse, he had to convince her to stop for a moment. This generally meant moving suddenly into her path to cut her off, forcing her to stand for him, or walking backwards to keep his hold on the teat while she pressed ahead. Often he kept company with other mares in the group who kindly looked after him while his mother foraged elsewhere. Between basic body maintenance and the enormous caloric drain of lactation, the bay mare had to eat almost constantly, going out of her way to select forage with the greatest nutritional benefit. The harem stallion, a powerful steel gray with bright dapples, was powerless to keep her in line, despite his large size and bossy attitude. Two years before, he claimed her as an adolescent filly and spent that summer driving her around and monitoring her every move. Since then, he had acquired three additional mates, cooperative mares who stayed together in a manageable group. None of them had foals, so they remained in reasonably good condition despite competing with other bands to graze on the lawn of the Dungeness ferry dock complex.

When rainfall is abundant, Cumberland Island blooms. Healthy-looking horses graze lush lawns with evident contentment. Foal survival improves, fertility increases, and the herd expands. When rainfall is scarce for long periods, upland grass withers, and ravenous horses crop it short. Marsh grasses, stressed by rising salinity in drought-stricken estuaries, are further pressured by overgrazing. Unable to meet their caloric requirements, horses burn fat and muscle for fuel,

In 2011, this 5-year-old mare grazed almost constantly to meet the caloric requirements of lactation, yet remained bone thin. The wild herd of Cumberland Island had reached the carrying capacity of the island's forage resources while rainfall was adequate. When years of drought reduced the vegetation, lactating mares had trouble finding enough to eat.

and become weakened by infections and parasite infestations. The most vulnerable members of the herd—foals, the elderly, and lactating mares—suffer most. If drought persists, many will die from starvation and illness.

For the bay mare, the need for social interaction was secondary to her relentless appetite. She knew when her favorite food, saltmarsh cordgrass (*Spartina alterniflora*), was exposed by the ebbing tide, and she made sure she was there to devour it, regardless of the activities of her herdmates. It was a daily challenge for her to consume enough calories to sustain herself and her foal.

Unlike the other East Coast islands roamed by wild horses, Cumberland is mostly dense subtropical forest of saw palmetto, magnolia, cedar, holly, pine, and myrtle. Majestic 300-year-old live oaks stand like wizened sentries along the pathways, Spanish moss trailing from their limbs.

One of the largest undeveloped barrier islands in the world, Cumberland is almost 3 mi/4.8 km wide at one point and nearly 18 mi/30 km long. The island is a mosaic of diverse ecosystems including

In 2009, this same mare was a healthy 3-year-old, newly claimed by the gray stallion as the start of his harem. It was an exceptionally wet year, and grass was abundant. Note the stallion's "snaking" posture as he drives her across the lush lawn.

primary and secondary dunes, interdune meadows, wax myrtle thickets, mowed lawns, fresh and salt marshes, and pristine beaches, as well as one of the largest maritime forests in the United States. More than 15 mi^2/40 km^2 are federally designated wilderness.

John F. Kennedy, Jr., brought Cumberland Island to national attention in the fall of 1996 when he married Carolyn Bessette in a quiet ceremony in the humble First African Baptist Church, then retreated to the privacy of Greyfield Inn. Cumberland Island was one of the few places where such a wedding could be conducted in utter seclusion. The event was coordinated by Carnegie descendant Janet "Gogo" Ferguson, part-owner of Greyfield, who developed a friendship with Kennedy during his regular visits to the island.

A few tracts remain privately owned, but most of the island is now a national seashore. Greyfield is situated in a private compound including more than 200 acres/81 ha of mostly undeveloped oceanfront property. The elegant mansion was built in 1900 for Margaret Ricketson, the daughter of Lucy and Thomas Carnegie and niece of Andrew Carnegie. Since 1962, the property has operated as an inn to generate income for its own upkeep.

After being twice rebuffed for trying to nurse, the foal approaches a third time, hesitancy showing in his expression and the set of his ears. This time the mare allowed him to feed, but she continued to tear at the tender *Spartina* revealed by the ebbing tide near Dungeness Dock. Without the pressures of overgrazing, these wispy sprouts could grow into a dense and healthy salt marsh.

A century ago, Cumberland Island was well known as a playground for the wealthy and powerful, but it lapsed into obscurity. Aside from nature lovers and birders, few people outside Georgia had heard of it before the Kennedy wedding thrust it back into the headlines.

Like the rope in a tug-of-war, this idyllic island is wrenched by conflicting laws, policies, and philosophies. The Wilderness Act (1964) demands that the National Park Service preserve much of the island without roads, structures, motorized equipment, or mechanical transport; yet within this wilderness, some can legally drive cars and live in houses. The National Historic Preservation Act (1966) calls for the protection of historic structures and objects in or near the wilderness area, but the Park Service was been obliged to maintain them without the benefit of trucks or power tools. Various laws protect legal contracts for real estate and reserved rights of use, but the rights upheld by these contracts frequently conflict with federal policies. The Park Service struggles to balance these competing interests, but it seems that every objective is at odds with another.

While the bay mare forages alone at the dock, the well-nourished gray stallion (right) waters at the artesian well with the rest of his band. Cumberland mares that are neither pregnant nor lactating are frequently underweight, but it is the nursing mothers that walk the narrow edge between survival and starvation.

Each of Cumberland's resources has its own advocacy groups, and the agendas of these organizations frequently collide. Impassioned environmentalists oppose current and former landowners, federal agencies contend with politicians, wilderness activists argue with historic preservationists, and horse-lovers clash with birders. Their only common ground is a love for the island and a desire to act in its best interest.

Popular legend holds that the wild horses of Cumberland Island descend from Spanish Jennets imported and maintained there between 1566 and 1675, when Spain had a fort and missions on the island. It is possible that Spanish horses were allowed to range freely to multiply and forage for themselves. Some speculate that the Spanish abandoned the island and left livestock behind, perhaps leaving horses too old, lame, or intractable to be useful to the Spanish elsewhere. Though this origin is possible, there is no evidence.

English imports probably came to Cumberland Island with James Oglethorpe in the late summer of 1739. He wrote, "I left a stud of the Trust's horses and mares when I went last for England, and the colts bred out of them are very good" (Bullard, 2003, p. 29). Two years later, Benjamin Martyn wrote that on Cumberland and Amelia islands "a stud of Horses and Mares . . . are bred without any expense" (Bullard, 2003 p. 29).

As on other islands, stockmen frequently released horses, cattle and other livestock into "feeding marshes" and onto islands to make use of otherwise unusable land and to contain the animals without use of fences These animals, sometimes subject to periodic roundups and at other times ignored for decades, often became wild.

Free-roaming horses were first documented on Cumberland Island in a 1788 letter to Edward Rutledge from Phineas Miller, but we do not know whether any descendants of the early lines remain. In a 1789 letter to Captain John McIntosh, a Cumberland wrangler arranged shipment on a flat boat for 13 young horses gathered from the island, "much better ones than the last" (Bullard, 2003, p. 70). Some of the remainder were probably eaten by 900 starving soldiers and freedmen during this time.

After the Civil War, Cumberland residents used most of the island for grazing free-range livestock. Records indicate that semi-gentled horses were swum from Cumberland Island to be sold on the mainland in 1866 much as the Chincoteague Ponies are penned today. This gather and swim might have been an annual activity, the drive probably starting in the marsh south of the Plum Orchard dock. At slack low tide, wranglers would press the horses to swim the Brick Hill River, then to the eastern shore of the shell-bottomed Cumberland River, to land at a point about 300 yd/274 m south of Cabin Bluff.

When the Carnegies came to the island in the 1880s, they stabled a number of well-bred domestic horses in their compounds. Plum Orchard had an impressive 15-stall stable filled with high-quality animals, and, farther south, the stable at Dungeness accommodated 60. The Georgia State Archives documents many horses that the Carnegies bought or sold on Cumberland Island and the presence of feral herds that ran at liberty, like the cattle and hogs, all over the island. The residents held wild horse roundups and "rodeos" to break them to saddle.

In the 1800s, free-roaming Cumberland Horses were popularly classified as Marsh Tackies. They were small in stature, and had the reputation of being exceptionally strong and rugged. At one point the horses might have carried a high proportion of colonial

The horses of Cumberland Island descend from a number of breeds, including the Tennessee Walking Horse, Thoroughbred, Paso Fino, Quarter Horse, Appaloosa, Spanish Colonial Horse, and Arabian.

Spanish blood, but over time outcrossing changed the character of the herd.

Many breeds have contributed genes to the free-roaming herd. Through much of the 19th century, visitors to the Robert Stafford plantation captured free-roaming "marsh tackies" and bought them for $5 each. It appears that Stafford imported foreign horses in the 1850s and maintained about 30–40 horses on the plantation through 1877. Thomas Carnegie purchased the Stafford and Dungeness plantations in 1881 and proceeded to improve the free-roaming herds by adding purebred horses of various breeds to the mix. From the 1880s through the 1960s, the Carnegie family released many horses—Appaloosas, Tennessee Walking Horses, Thoroughbreds, even retired circus horses. In 1896, the Carnegies bought a white stallion from the stud farm of Czar Nicholas II. In 1921, a train-car load of beautiful Arizona mustang mares probably brought an infusion of Spanish blood. It remains unclear which, if any, of these horses contributed to the modern free-roaming herd. In the 1950s, Lucy Ferguson and

A thin black mare with the conformation of a Tennessee Walking Horse grazes under the moss-draped live oaks at Plum Orchard. After the author took this photograph, the mare's yearling son emerged from a copse of cedars, shedding light on the reason for her emaciation. When resources are scarce, lactating mares become vulnerable to malnutrition. A well-nourished gray stallion with Appaloosa characteristics forages in the distance.

her staff castrated every wild male foal that they could capture, then released their own fine stallions to breed with the wild mares. As recently as 1992, a resident added four registered Arabians to the wild herd to improve conformation.

In 1991, Goodloe et al. reported that the Cumberland Horses are genetically similar to other East Coast island populations, especially the Assateague Ponies. Assateague Ponies carry mostly old Spanish, mustang, and Arabian bloodlines, with Shetland Ponies introduced in the 1920s to add lively markings to the herd. The horses of the North Carolina Outer Banks are a unique, mostly homogeneous strain of Colonial Spanish horse, apparently descended from old Spanish bloodlines.

The horses of Cumberland also carry genetic markers from American breeds such as the American Thoroughbred, the Tennessee Walking Horse, and the American Quarter Horse. These breeds descended largely from a handful of Spanish, English, Dutch, French,

On a visit to Cumberland Island in the 1990s, the author photographed an exquisite white Arabian mare running wild in a group of roans and Appaloosas near the Dungeness ruins.

Arabian, and Barb horses imported into the colonies in the 17th and 18th centuries. Consequently, each of the American breeds share so many genetic markers with one another that pinpointing lineage can be difficult or impossible. Arabian and Paso Fino horses have also made their mark on the genome of the Cumberland herd. It is probable that the horses currently on the island descended primarily from stock associated with the Carnegie estates, which probably carried these bloodlines.

Compared to other East Coast island horses, members of the Cumberland herd are generally taller, longer-legged, and sleeker. Whereas Assateague Ponies are mostly pinto or sorrel, Cumberland Horses range from bays and chestnuts to grays, blacks, browns, and roans. Many horses have white markings such as blazes, forehead stars, and socks. Some horses have Appaloosa patterns—such as roan with white blankets, dark with white "snowflake" markings, or spotted blankets across the hindquarters—and show other Appaloosa traits such as a sparse mane and tail, striped hooves, mottled skin, and white scleras.

Horses have been in continuous residence on Cumberland Island for hundreds of years, but exactly how long is anyone's guess. Over the last 500 years, the island has been repeatedly developed and exploited, then abandoned, and many of these contingents brought horses with them. Native Americans made minimal impact, but the Spanish, English, and American residents transformed the island with forts, missions, logging, cotton plantations, estates, a hotel complex, and numerous other projects. Between waves of human habitation, mansions crumbled and native vegetation and wildlife encroached. Since the Cumberland Island National Seashore was established in 1972, and especially since Congress designated 9,886 acres/4,001 ha (more than half the seashore) wilderness in 1982, the island's primitive persona has been encouraged to reassert itself in many places. Up to 10,731 acres/4,343 ha may be added to the wilderness area in the future.

Paleo-Indians arrived in the Florida region as early as 13,000 years ago, and it is reasonable to think that these early people inhabited the area that would become Georgia at least as early. It is likely that eastern North America was peopled long before this date, but evidence of this habitation is scarce. Between 18,000 and 12,000 years ago, the shoreline hugged the edge of the continental shelf, about 70 mi/113 km east of the present beachfront. When people first made their homes in the area, it was far from the ocean, and they began foraging on the present island more than 4,000 years ago.

At the time of European contact, the indigenous Timucua Indians burned land for agriculture and to create a favorable environment for the white-tailed deer. They subsisted largely on seafood and wild game and left behind ceramic shards and numerous shell middens throughout Cumberland Island. Many roads and trails on the island today follow their ancient footpaths. They spoke a language apparently unrelated to any other. Although the Timucua survived until the early 19th century, and Spanish missionaries preserved some of their language in writing, little else can be said with confidence. Those who lived on Cumberland Island and the nearby mainland are sometimes called, and may have called themselves or their village, Tacatacuru. But Milanich points out,

> Just how the natives did refer to themselves is something of a puzzle. Evidence exists that when asked by a Frenchman or a Spaniard what a place was and who its occupants were, villagers responded, "this is our land" and "we are us." (1996, p. 45)

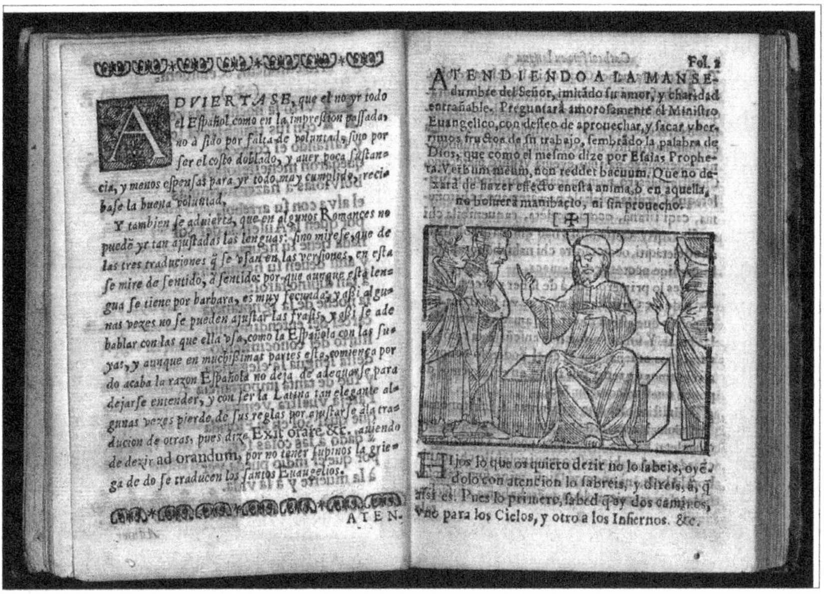

Spanish missionaries spread the Catholic faith to native tribes in what is now North Florida and Georgia by teaching from a catechism printed in both Spanish and Timucua (Pareja, 1627). When Europeans first made contact in the 16th century, the Timucua may have numbered as many as 200,000. Three hundred years later, they were extinct, victims of warfare, bondage, and European diseases. Courtesy of the Jay I. Kislak Collection, Rare Book and Special Collections Division, Library of Congress.

The first European to visit Cumberland was probably Jean Ribault, a Frenchman, in 1562. The Spanish named the island San Pedro, and in 1566 the Spanish governor of Florida directed construction of a fort on the island, also named San Pedro. The governor then established the Franciscan missions of San Pedro de Mocama and San Pedro y San Pablo de Puritiba (Puturiba, Potohiriba . . .) to convert the natives to Christianity. Researchers believe that missionaries brought livestock to the island. The first mission seems to have lasted until about 1660. The second was apparently active only in the mid-1590s. A third, San Felipe, lasted till around 1684. After attacks by pirates in 1683 and 1684, missionaries and natives alike seem to have fled to St. Augustine. The Yamasee from farther north had settled the island by the time Oglethorpe arrived.

In 1736, Oglethorpe, founder of the English colony of Georgia, established two forts on the island—Fort St. Andrew at the north end and Fort Prince William to the south. Cumberland's present name came about through the suggestion of an Indian boy named Toonahowie, who visited London with his uncle Chief Tomochichi and General Oglethorpe. Toonahowie struck up a friendship with the 13-year-old son of King George II, Prince William Augustus, Duke of Cumberland. To seal their friendship, on parting William gave Toonahowie a gold watch. Toonahowie, in turn, asked General Oglethorpe to name this island after the Duke of Cumberland. Fort Prince William, a substantial structure built on South Point to accommodate 200 men, also honored the young duke. Dungeness, originally constructed as Oglethorpe's hunting lodge, was named after a headland in Kent.

Spanish forces clashed unsuccessfully with British soldiers, many of them Scottish Highlanders, at the southern fort in 1742, and their defeat ended Spanish incursions into English Georgia. After the battle, however, the English believed the southern coast of Georgia to be indefensible against Spanish, French, and Indian raids. The forts, constructed of timber and sand, were reclaimed by the elements; no trace remains today. From 1748 through the 1750s, the island was inhabited mostly by "bandits, pirates and other lawbreakers" from the English and Spanish colonies (Dilsaver, 2004, p. 23).

In 1763, the Treaty of Paris ended the French and Indian War (Seven Years' War) and granted the English rights to Florida, removing Cumberland Island from the frontier and making land grants there desirable. Before the American Revolution, there were few settlers on Cumberland, but thereafter planters moved in to grow rice, indigo, and corn and raise horses, cattle, and hogs. Colonists harvested Cumberland's live oak and cedar for ships' timber, lumber, and shingles.

After the war, General Nathanael Greene, commander of American forces in the south, acquired Mulberry Grove plantation near Savannah. He later bought extensive property on Cumberland Island, intending to harvest live oak to sell for shipbuilding and to build a home for his family on the site where Oglethorpe had maintained his hunting lodge, Dungeness. Unfortunately, Greene died before he could begin to build or turn a profit. Greene wrote that in 1785, shortly after the war, at least 200 horses and mules roamed the island. Goodloe et al. (1991) offer corroboration, a 1788 letter from island resident Phineas Miller to Edward Rutledge that mentioned free-roaming horses.

In 1796, Greene's widow, Catharine, married Phineas Miller, the general's former secretary and her children's tutor. They built a four-story mansion, also named Dungeness, surrounded by 12 acres/5 ha of terraced gardens that grew many exotic foods. The Millers frequently held elegant parties, and Cumberland became a social hub for the affluent. Phineas Miller also died, leaving Catharine once again alone. In the years to come, the pattern would repeat—men would die young, leaving Cumberland Island molded and controlled by strong, intelligent widows.

When British Admiral George Cockburn briefly occupied Cumberland, he and his men used Dungeness as their headquarters and hospital. To produce social turbulence, Cockburn freed the slaves there and so attracted many runaways from around the region, who were ultimately transplanted to Trinidad.

Another Revolutionary hero, General Henry "Light Horse Harry" Lee, III, suffered internal injuries while defending a friend and freedom of the press in Baltimore, went to the West Indies to recover, and retreated to Cumberland Island as his health declined. He died on Cumberland in 1818 in the home of Louisa Shaw, the daughter of his former commander, Nathanael Greene. He was later exhumed and reinterred beside his son, Robert E. Lee, in Lexington, Va.

From the end of the Revolution to the Civil War, Sea Island cotton—a variety that commanded higher prices by virtue of its unusually long fibers, was the major cash crop of Cumberland Island. More than half of the island was cleared for farming. Slaves lived in small cabins. Today a cluster of stone chimneys marks their locations. During the heyday of cotton plantations, Cumberland supported a population of 65 whites and 455 slaves. Other products cultivated included timber, citrus fruit, figs, dates, olives, and pomegranates. Horses were necessary for working these plantations and for traversing the 18-mi/30-km length of the island.

When the Civil War began, the planters of Cumberland retreated to the mainland for safety. They returned to find Dungeness burned to the ground except its stone walls and chimneys.

Robert Stafford, Jr., son of Thomas and nephew of Robert Stafford, acquired land over time until his holdings totaled 8,125 acres/3,288 ha, located around the present-day Sea Camp area. (At the time, the terms *junior* and *senior* were often used to distinguish between two

A collection of skulls, many of them from horses, decorates the tabby wall surrounding the Stafford mansion.

men with the same name, lineal relatives or not.) Census reports show that Stafford kept 30–40 horses on his plantation between 1850 and 1870 despite the fact that most Cumberland Island horses were sold to help the Confederate effort during the Civil War. From 1800 to 1880, the Robert Stafford plantation sold Marsh Tackies, ponies captured from free-roaming island stock, for $5 each.

Stafford never married, but fathered 6 children with a slave named Elizabeth. During the 1850s Stafford owned up to 348 slaves, who were emancipated when he was in his 70s. The profitable island plantations owed their success to slave labor, and after the Civil War, agriculture was no longer sustainable. Before the Civil War, there were 10–12 large, productive plantations on Cumberland, but by 1876, not one acre was in cultivation.

Rather than leave his land and see it divided among his freed slaves as abandoned property, Stafford remained until he died in 1877, destitute and bitter about his change of fortune. After his death, his nephews inherited his holdings. The six biracial children he fathered with Elizabeth did not receive any land on Cumberland.

Stafford's story was not unique. After 1865, there was a reshuffling of the social order, and many prominent landowners faced poverty. Planters could not afford to hire laborers or resisted employing former slaves. Some former masters evicted the freedmen and burned their cabins.

By 1878, two hotels were operating on the northern end of Cumberland Island in an area known as High Point. They reached a peak in operations in the 1890s and 1900s and shut down by 1920. This property later passed to the Candler family of Atlanta, whose wealth came from the astonishingly popular beverage Coca-Cola.

Thomas Morrison Carnegie, brother and partner of steel magnate Andrew Carnegie, bought the plantation at Dungeness in 1881 for his wife, Lucy, and their family. They built another Dungeness mansion on the site of the original, graced by verandas, turrets, and gables and boasting 50 rooms. Dungeness was evidently built on a prehistoric shell midden, which undoubtedly provided the material for the 6-inch-thick walls of the tabby structure.

Like the earlier Dungeness, this great house soon became a social center for the wealthy. Thomas and Lucy entertained guests with activities such as hunting, fishing, golfing, and cruising aboard their yacht, also named *Dungeness*.

Thomas Carnegie died in 1886 at the age of 43, leaving Lucy with nine children and a large inheritance. Lucy went on to acquire 90% of Cumberland Island, turning it into a vast, self-sufficient family preserve staffed by about 200 employees. She set about expanding the Dungeness complex to include more than 20 buildings, as well as walls, decorations, and a pergola (colonnaded walkway that supports climbing plants). A recreation and guest house east of the mansion included a heated pool, a steam room, a recreation room, a squash court, and guest bedrooms. Other houses, docks, and structures were built all over the island. Seven of Lucy's nine children married, and she presented four of them with mansions on Cumberland Island.

The Carnegie estate maintained an ornamental upper garden while the lower garden became a subsistence garden of vegetables and fruits. The complex included greenhouses to grow cut flowers, an olive orchard, crops such as prize-winning Sea Island cotton, and livestock such as beef and dairy cattle, pigs, and poultry. Carriage and pleasure horses roamed free on the island, and about 60 were kept stabled at Dungeness.

Lucy bestowed the Stafford mansion on her eldest son, William, along with cash for renovations; but in 1900 the old

The third Dungeness, still impressive despite decades of neglect, as photographed for the Historic American Buildings Survey the year before vandals burned it down. Photograph by P.E. Gardner (1958), courtesy of the Prints and Photographs Division, Library of Congress.

homestead burned. On the site, William built a second Stafford house, which was similar to the previous structure, but with a fire-resistant stucco exterior.

Lucy's wedding gift to her son Thomas Carnegie II was a 29-room house known as The Cottage, built on the grounds of Dungeness. It burned in the 1940s. Lucy's daughter Margaret (nicknamed Retta) used her husband's wealth to build Greyfield, a two-story house 2 mi/3.2 km north of Dungeness. Greyfield now operates as an elite and expensive inn. In 2013, the most economical room was $425 a night off-season, single occupancy, for a minimum of two nights (three on holidays). Air conditioning, meals, and access by private ferry were included; telephones, television, and Internet access are not.

Lucy Carnegie built Plum Orchard for her son George when he married Margaret Thaw in 1898. Eight years later, Margaret expanded the mansion to include 30 principal rooms, 12 bathrooms, and many smaller rooms.

Lucy set up a trust that protected the family's holdings until her last child died in 1962. She ensured that she would maintain control of Cumberland Island and that she would not lose her fortune to any irresponsible actions of her children. Her six sons lived a life of leisure, and none of them ever held a job.

When Lucy died in 1916, Andrew II, Thomas II, and Margaret remained trustees of the estate. Initially, the Carnegies maintained their affluent lifestyle, but finances grew tight in the mid-1920s and tighter with the stock market crash of 1929. Although the Carnegies valued the primitive nature of the island, financial difficulties forced the family to generate income by closing Dungeness and capitalizing on Cumberland's resources. They considered raising cattle and hogs, lumbering, sawmilling, planting, shrimping, operating freight and passenger boats, and investing in developments such as hotels and amusements. In the 1920s, A.A. Ainsworth of New York almost purchased Cumberland to found a major development similar to Coral Gables, Florida, but the project never materialized.

In the 1950s, several companies bid to strip-mine titanium-bearing ilmenite from 7,000 acres/2,800 ha of Cumberland Island, an undertaking that would have destroyed vegetation and other surface features of the island. Luckily, legal difficulties and a sudden drop in the price of titanium saved Cumberland from disembowelment.

By the 1950s, the gardens at Dungeness were no longer maintained. In 1952 Robert Ferguson sought bids to demolish Dungeness, but did not follow through on razing it.

Around that time, Cumberland Island was subject to poaching and vandalism, and Lucy Ferguson and the other landowners resorted to "hunting" the poachers to repel them from the island. Employee J.B. Peebles shot at one such poacher in 1959, and shortly afterward, Dungeness was destroyed by a fire that could be seen for miles along the mainland coast. Several Florida poachers were the likely culprits, but nobody was ever arrested for the crime. The ruins of Dungeness remain a popular feature of Cumberland to this day, and wild horses favor grazing on the grassy grounds.

In 1951, Lucy Ferguson and her husband turned to cattle ranching. Their cattle opened the forests, destabilized sand dunes, trampled marsh, and altered the population dynamics of deer and horses so that it is difficult to assess the true carrying capacity of the island today.

This pronghorn (top right, beside a white tailed deer) and elk (bottom) are native North American wildlife, but when brought to Cumberland Island in the late 19th century as a novelty, they became unequivocally exotic species in this environment. More enigmatic is the status of horses as native or exotic wildlife within the barrier island environment. If native is defined as present at the arrival of the first Europeans, horses are clearly exotic; but if defined as present at the arrival of people, horses are a native species, and our Pleistocene ancestors probably contributed to their extinction. Though horses are native to East Coast salt marshes, there is no evidence that they lived on barrier islands before reintroduction. Ocean levels were about 400 ft/122 m lower than they are at present, and what is now submerged continental shelf was exposed as much as 60–90 mi/97–145 km east of the present shoreline (Carroll, Kapeluck, Harper, & Van Lear, 2002). Cumberland Island Ricketson Family Photograh Album *ca.* 1898–1900. Courtesy of the Lucy Coleman and Thomas M. Carnegie Family Papers, ac 1994-0003M, Georgia Archives.

The Carnegie heirs considered options that would maintain the island's wild beauty while allowing them to maintain a presence. In 1955, the Park Service published a report titled *Our Vanishing Shoreline*, which expressed concern that the wild coast of the United States was being swallowed by developers (USNPS, 1955). Of the 3,700 mi/6,000 km of coast, only 6.5% was preserved for public recreation, and most of that was around Cape Hatteras. The Park Service sought to acquire 15 additional sites and considered Cumberland Island ideal because it contained "practically all the desirable features for public enjoyment" (Dilsaver, 2004, p. 80).

Carnegie landowners invited the Park Service to consider Cumberland as a potential national park, and it eagerly agreed. Secretary of the Interior Stewart Udall promised that if the Park Service acquired Cumberland Island, it would maintain the property largely undeveloped, preserving the unspoiled beauty that the Carnegies had cherished. The Park Service offered to obtain land over time, paying the residents substantial sums but allowing them to live and drive vehicles on the island.

At about the same time, the state of Georgia was eyeing Cumberland Island as a state park with the intention of developing it into a resort "for the average man" (Dilsaver, 2004, p. 81). A bridge would connect Cumberland to the mainland, roads would be built to and on the island, and the state would "propagate it as a vacationland" (Dilsaver, p. 83).

At the time, land owners had differing visions for the future of the island. In 1968, while the Park Service negotiated for a national seashore, three of the Carnegie landholders sold large tracts totaling 3,000 acres/1,200 ha, or one fifth of the total Carnegie holdings, to Charles Fraser of the Sea Pines Development Company. Fraser was the developer and major owner of a large resort complex on Hilton Head Island, South Carolina.

Fraser envisioned Cumberland as a residential retreat for the affluent with about 150 homes; golf courses, natural areas, and recreational facilities and a fire department, airport, and medical center. He even incorporated the Park Service into his plans, supporting a small national park within his resort community. As relations soured, Fraser set out to gain control of the island by any means, intending to push the landowners off their property

The multi-tiered lawns surrounding the ruins of Dungeness are the favored grazing areas for several bands of horses.

and block their influence. Fortunately, he failed, and Cumberland escaped development.

In 1961, the U.S. Air Force and the National Aeronautics and Space Administration sought an East Coast launch site for the developing space program. They winnowed the list of potential properties to two, Cumberland Island and Cape Canaveral. NASA actually preferred Cumberland for this purpose, but Cape Canaveral was chosen because of its proximity to Patrick Air Force Base. If Cumberland had been acquired by NASA, neither the Carnegies nor the Park Service would have had use of it.

The remaining Carnegie and Candler descendants, who together held 85% of the island, were disturbed by the prospect of outsiders transforming an island that had been in their families for decades. Together with a group of conservationists, they succeeded in blocking development and establishing a national seashore on Cumberland Island while retaining their homes.

In the early 1960s, Little Cumberland Island, at just north of the main island, changed hands twice and was bought by conservation-minded investors who planned to develop 200–300 residences, leaving the rest of the island in a natural state.

In 1971 the Carnegie heirs sold or donated most of their property to the Park Service. On October 23, 1972, President Nixon signed legislation to establish Cumberland Island as a 40,500-acre/16,400-ha national seashore, making it one of the largest mostly undeveloped barrier island preserves in the world. Several parcels of private property or retained-rights lands escaped Park Service control. There were 18 retained-rights lots within the seashore boundaries deeded to allow residents to live on them for periods ranging from decades to life, with the Park service to assume ownership of the parcels when the agreements expired. Some residents rent out their properties, some use them seasonally, and at least one lives on the island year-'round.

When the Park Service acquired Cumberland Island, its original plan was to create a recreational park that would attract 10,000 visitors a day with numerous interpretive sites and activities. Environmental groups, residents, and the public argued that the park should preserve the island's unique wild character, and the Park Service agreed to a primitive park with a 300-person daily limit and minimal public facilities. The *Cumberland Queen*, operated by Lang's Seafood, the concessionaire for the park, ferries people to and from the island, taking up to 150 per trip.

Cumberland Island is not far offshore. At the Cumberland Dividings, between Crooked River and Black Point, the distance from high land on the mainland to high land on the island is only about 300 yards/274 m. Marshes block direct east-west passage for boats, however. To venture from the port of Saint Marys to a landing on the island, they must travel up to 6 mi/10 km.

The first Europeans to claim Cumberland Island found large stands of live oak and pine with less understory of saw palmettos (*Serenoa repens*) and other ground cover than exists today because of centuries of intentional burning by the native people. The forest remains dense despite extensive British and American harvesting of live oak for shipbuilding.

The marshes, beaches, meadows, forest, and ocean habitats teem with wildlife. Tree frogs trill loudly for much of the night, their throats ballooning like bubble gum, at daybreak sleeping off the evening's

Sea turtles have been making a comeback on Cumberland Island, largely because of aggressive conservation efforts. Wildlife biologist Doug Hoffman demonstrates the effectiveness of a barrier that prevents raccoons and hogs from preying on nests. The year 2010 saw a record 486 successful nests.

debauchery camouflaged on palmetto leaves. Log-like alligators (*Alligator mississippiensis*) lurk half-submerged in the freshwater shallows, and river otters (*Lontra canadensis*) cavort in their quest for fish. Throughout the moist, steamy woods, armadillos (*Dasypus novemcinctus*) root for grubs in rich humus. White-tailed deer (*Odocoileus virginianus*), smaller than in northern forests, venture onto the meadows as the sun drops behind the horizon. Venomous snakes such as diamondback rattlesnakes (*Crotalus adamanteus*), timber rattlesnakes (*Crotalus horridus*), and cottonmouths (*Agkistrodon piscivorus*) hunt scampering rodents.

The waters off Cumberland are critical breeding and calving grounds for endangered right whales (*Eubalaena glacialis*). Manatees (*Trichechus manatus*), up to 15 ft/4.6 m long and weighing as much as 3,000 lb/1,361 kg, graze on *Spartina* and algae in salt-marsh creeks. Groups of as many as seven manatees commonly loiter in the salty water around the ferry docks at low tide, drinking the freshwater runoff while people wash their boats.

Sea turtles nest on the wild beaches. These reptiles were hunted on the island through the early 1900s. At one point, the Carnegies preferred turtle eggs to hen's eggs and enjoyed green turtle soup. Eggs were found by thrusting a sharp stick into the sand of a potential site. If it was a nest, the stick would withdraw covered in sticky raw egg. Over the past century, once-abundant marine turtles became rare on Cumberland and elsewhere.

They are, however, apparently making a comeback, with Hoffman counting 336 nests in 2006, 252 in 2009, and 486 in 2010, 700 nests in 2012, and 561 in 2013. "As a whole, the state of Georgia has recorded its highest nest totals during the four nesting seasons in years 2010–2013," says Hoffman (personal communication, July 20, 2014). Interns from the Student Conservation Association—volunteers that provide manpower for Park projects—document the location of the nests, secure them against predators, and give educational talks to visitors.

Migratory birds use the island as a stopover. Cumberland Island is designated a United Nations International Biosphere Reserve because it offers critical habitat for endangered wildlife such as least terns (*Sternula antillarum*) and loggerhead sea turtles (*Caretta caretta*).

More than 323 species of birds are residents or seasonal visitors, including threatened and endangered species such as bald eagles, least terns, Wilson's and piping plovers, and wood storks. "We have some of the highest shorebird numbers in the state, both during nesting season and the winter migratory period," says Hoffman.

> Primary beach/dune nesters in the spring/summer include Wilson's Plovers, American Oystercatchers, Least Terns, and Willets. Our wading-type birds—wood storks, herons, and egrets—do not nest on the island, and we presume that the nesting habitat has been lost or severely decreased due to prolonged droughts and the absence of wildfires that would have kept freshwater wetlands free from woody vegetation encroachment (D. Hoffman, personal communication, March 18, 2011).

Black bears (*Ursus americanus*) and endangered Florida panthers (*Puma concolor*) were native to the island, but were hunted to extinction over the last two centuries. Bobcats (*Lynx rufus*), a species that vanished from Cumberland in 1907, either from human activities or from natural causes, have been reintroduced.

Researchers from the University of Georgia released fourteen adult bobcats in the fall of 1988 and 18 more in fall of 1989. They thrived on the island, preying primarily on marsh rabbits but rounding out their diets with cotton rats, grey squirrels, raccoons, birds, cotton mice, white-tailed deer and young hogs. To monitor the genetic health of the felines, in the winter of 2011–2012 researchers sampled DNA from their scat and compared it to genetic samples from the founding population.

Destructive, elusive, and prolific feral hogs are perpetually vexing to the Park Service. They are adept at evading capture and multiply rapidly, with large litters, short gestation periods, and early puberty. Hogs can breed as early as 6 months of age, but more typically do not start to reproduce until 8–10 months. The usual litter size is 4–6 piglets, and they can produce up to two litters annually, though one a year is typical. They live in family groups called sounders, which usually include two sows and their offspring. Adult boars live alone and join a group only to mate.

Their preferred habitats are palmetto forest and salt marshes. Feral hogs will eat anything from grasses, fruit, nuts, tubers, and seeds to rodents, frogs, fawns, and carrion. The hogs stand accused of consuming the eggs of ground-nesting birds and sea turtles, uprooting and destroying endangered plants, and competing with native animals for resources.

The Park Service instituted a program of live-trapping and removal in 1975, and by 1979, 1,100 feral hogs had been removed from Cumberland. The designation of nearly 10,000 acres of Wilderness in 1983 significantly impacted the NPS' ability to manage the feral hog population by closing roads that were used for trapping. While hog control efforts continued at a reduced level, large scale hog management efforts did not resume again until 2000. Drought and hunting pressures have reduced the hog population considerably. In 2009, 465 hogs were harvested; in 2010, 356; in 2011, 233; in 2012, 221; and in 2013, 328. Hoffman estimates the 2014 hog census at fewer than 200 animals.

Not as charismatic as horses and usually more numerous, feral swine are seen universally as a nuisance animal at Cumberland and on many other federal lands. With all of the fervently negative sentiment, it comes as a surprise that some research portrays the hogs (in reasonable numbers) as a relatively benign presence. Jill Baron (1982) found that the feral hogs of Horn Island, Mississippi, ignored the sea oats and other vulnerable plants to scavenge tideline trash, dead fish and crabs, and other carrion. Disturbances created by the rooting of hogs recovered within months, sometimes with greater plant cover. After 150 years of hog habitation, the vegetation of Horn Island was indistinguishable from that of a swine-free island nearby.

The feral hogs of the Sea Islands may be longer-established than the horses. A study of mitochondrial DNA from feral hogs on Ossabaw Island (near Savannah) suggests origins in the Canary Islands from Spanish stock. The American Livestock Breeds Conservancy (2009) puts them on its critical list. On Cumberland, this Spanish ancestry is more dilute. The first swine were probably brought to the island by Spanish colonists and missionaries in 1562, and they probably have been free-roaming since that time. Their bloodlines have blended with the genes of other domestic stock that escaped or was released over the centuries. In the 1940s and 1950s, lifetime resident Lucy Ferguson released a number of domestic swine to enlarge the gene pool of the feral population, and European wild boars may have been added to make the swine more desirable to hunters.

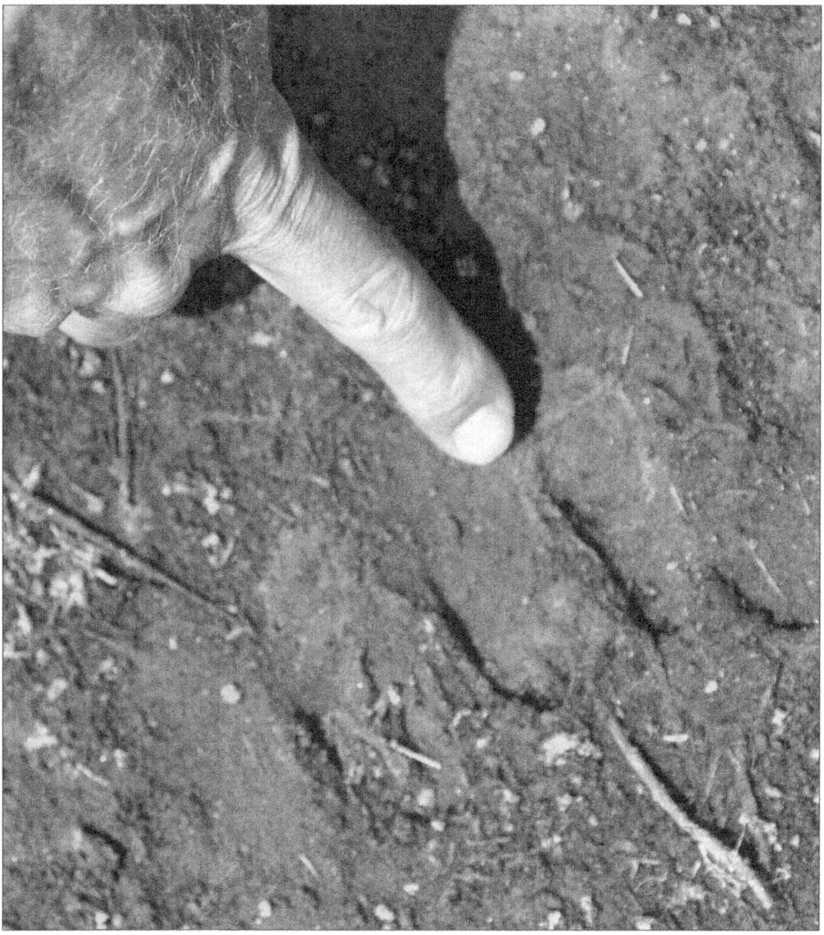

Hoffman explains that the track of a wild hog can be distinguished from that of a deer by the rounded tips of its cloven digits. As summer takes hold, Hoffman spends long nights on the beach hunting pigs that have the potential to consume sea turtle nests.

When the author toured Cumberland Island with Hoffman in 2011, more than 150 horses had maintained a relatively stable population, their numbers neither increasing nor decreasing significantly over the last few years. Hoffman said the lack of growth indicated that the herd had reached the carrying capacity of the island.

Carrying capacity concerns the number of animals that can be supported by available habitat. Over the long term, horse herds managed at carrying capacity overgraze certain species of plants and inflict damage that reduces the future carrying capacity of the habitat. The

Lactation is nutritionally demanding and can nearly double a mare's daily energy requirements. On average, mares produce about 3 U.S. gal./11.4 L of milk daily—approximately 450 U.S. gal./1,700 L of milk in the first five months of her foal's life. During a dry year like 2011 on Cumberland Island, it can be extremely challenging for a mare to support both herself and her foal.

system becomes less productive and less diverse, in effect taking a step backwards in evolution.

Moreover, carrying capacity naturally fluctuates with environmental conditions. For a few years prior to 2009, rainfall was intermittently abundant, allowing forage to grow robustly, and the herd to expand. Then, from 2010 into 2013, severe drought took hold, and forage grew sparse. Then the rains came with a vengeance, and the island became lush once again. "We have had so much rain since the late spring of 2013 that all freshwater wetlands and dune fields have essentially been flooded for the last 15 months, " said Hoffman. "Truly a boom or bust situation" (personal communication, July 21, 2014).

This waxing and waning of resources had little effect on the herd: the census was 143 in 2003, 137 in 2004, 139 in 2005, 143 in 2006, 120 in 2007, 124 in 2008, 131 in 2009, 120 in 2010, 148 in 2011, and 136 in 2012. Observers only counted 108 horses in 2013, apparently due to a series of severe thunderstorms which disrupted the census. In 2014,

During a wet year like 2009, the island is lush and verdant, and forage is abundant. This mare has remained in good flesh despite months of nursing her robust colt.

the count was consistent with that of previous years—143 horses were identified (D. Hoffman, personal communication, July20, 2014).

Park Service policy is to remove exotic animals that threaten the environment. But as we have seen, it also is mandated to preserve and interpret cultural resources. "The horses are a non-native species just like the free-ranging cows that were removed in the 1980s and the feral pigs that we manage 365 days a year," says Hoffman. "Yet public sentiment differs significantly for each species" (personal communication, March 18, 2011).

Free-roaming horses have become a symbol of Cumberland Island to many people. Hoffman admits,

> This is a tough one because of the emotions and wide range of opinions regarding horses on this island . . . and emotion tends to override fact when you're trying to get a message across. Opinions on this issue vary even among the park's staff. The park realizes that while the horses are not native to the island, they are a big part of the visitor attraction. Ideally, a plan should be developed that allows for management of the herd and protection of the island's natural resources, while

A wild horse grazes beneath the wing of a private airplane on the grounds of the Stafford plantation in late May, 2011. Even on the expansive lawn area, grass is dry and sparse.

> allowing for continued public enjoyment. (Personal communication, March 18, 2011)

On Cumberland Island, each band claims a home range of about 650–1,062 acres/263–430 ha. As in other locations, dominant stallions control harems, and dominant mares lead their bands to the best forage, the best water sources, and the best places to avoid insects.

Goodloe et al. (2000) found that most of the bands on Cumberland contained 2–10 horses, with an average number of 4–5. Almost 85% of bands were harems containing mares, youngsters, and one dominant stallion. About 10% of bands had more than one adult stallion, and mare and foal groupings comprised about 3.4% of bands. Dominant stallions typically maintained control of their harems for short intervals ranging from a few months to less than 3 years. Only 37% of dominant stallions retained harems over the duration of the 5-year study.

Foals are usually fathered by the dominant stallion within a mare's band. Stallions avoid mating with their own adolescent daughters, but in multi-stallion herds, the beta male sometimes mates with these fillies. If a number of mares come into heat simultaneously while two bands are in close proximity, sometimes a mare will mate with a rival stallion.

In this 2011 photograph, horses graze the sparse dry lawn surrounding Plum Orchard. The horse in the foreground, the adolescent son of the mare in background, is in excellent condition, having been sustained by her milk for more than a year. The mare, however, is extremely thin and appears to have difficulty recovering her body weight with the forage available to her.

Mares return to estrus 1–3 weeks after foaling, and most will conceive in this "foal heat." With an 11-month gestation (range of 320–365 days), it is possible for a mature mare to be pregnant nearly all the time, year after year, even while nursing foals. More commonly, mares will foal in alternate years or two years out of three.

Studies conducted on domestic horses demonstrate that about 80% of sexually active mares will achieve pregnancy over the course of a year, which, incidentally, is close to the 85% conception rate for women having unprotected intercourse for a year. Of these equine pregnancies, more than 12% miscarry, most commonly between days 25 and 31. Pregnancy loss is more common in lactating mares who conceive immediately postpartum and among adolescent mares.

Interestingly, a well-nourished, healthy mare is more likely to conceive a male foal, while a mare in poor condition is more likely to conceive a female, though at mid-gestation mare condition was not a significant predictor of offspring gender. Male foals are great caloric investment for the mare. Colts gestate longer than fillies and nurse

This leggy newborn on the Stafford Plantation has roughly a 68% chance of surviving her first year. High mortality rates are common among wild-born foals. The sparse, short-cropped grass foretells the difficulty his mother will have consuming enough calories to sustain him.

40% more often during their first two months, presumably to give them a competitive edge in their lives as stallions. In nutritionally poor years, however, newborn colts are more likely to die than newborn fillies.

Goodloe et al. (2000) studied the population dynamics of the Cumberland horses from 1986 to 1990 and saw the herd increase from about 186 to about 220 individuals. An estimated 61% of mares foaled each year. Of the 178 foals born during the study, 123 (69%) survived to their first annual census—61.1% of the fillies and 58.8% of the colts "during 1987 to 1989 (the 3 years of most intensive observations)" (Goodloe et al., 2000, p. 117). Among wild herds, high mortality for young foals is a natural mechanism that keeps the population in check.

The Cumberland herd does have a foal-loss rate considerably higher than the 20–25% average for North American wild horses, but it is not by any means the highest. The Kaimanawa wild horses of New Zealand average a 50% foal-loss rate, and the Nevada Great Basin herd saw foal mortality as high as 88% due to predation by mountain lions. By comparison, among humans the mortality rate for children

less than 1 year of age was up to 20–30% as recently as the 1800s, and it remains as high as 16% in Afghanistan.

From 2003 through 2011, about 14% of foals survived each year. Hoffman explained,

> Stories from island residents and park personnel suggest that foal mortality is common. The earliest foals are born during a time when there is normally not much in the way of grass from spring green-up. One would suspect that mares nursing during that time would not be in optimal condition to produce quantity/quality milk versus mares that began nursing in April after the green-up. Also, this island goes through long periods of drought conditions on occasion, further affecting the availability of food and mare condition/milk production. (Personal communication, March 18, 2011)

During Goodloe's five-year study period, the Cumberland herd increased by only 4.3% annually. Hoffman says that the mortality rate is very close to the birth rate, as evidenced by the fact that the population has remained stable over the last few years. Causes of mortality include high parasite loads, drought-related stress, age, coyote predation, accidents, and perhaps Eastern equine encephalitis and West Nile virus.

Unlike sheep, goats, or even cattle, horses gestate only one offspring at a time. Twin conceptions are uncommon in horses and usually abort well before term. If the pregnancy continues until viability, the foals usually die at or soon after birth, especially in a wild herd. Among domestic horses, only 14% of twin foals survive to 2 weeks of age. Typically, one or both are undersized and malnourished because of fetal competition for nutrients.

"The first foals are usually born in early to mid-March and most are born by the middle of April," says Hoffman (personal communication, March 18, 2011). Foals are active *in utero,* often moving 55 times an hour from the 5th to the 9th month of gestation, and up to 84 times an hour in the last few days before delivery. As the mare nears term, her abdomen visibly ripples with the movements of her growing fetus.

Days or weeks before the onset of labor, the mare's udder will swell and her teats will develop a waxy coating of honey-colored dried colostrum. The muscles around her tail head soften and sink. The lips of her vulva elongate.

As parturition approaches, the stallion may show increased sexual interest in the mare, courting her and herding her around. She rebuffs his advances, but he continues to pester her, pushing his nose under her tail and growing excited by the odor of the fluids. Sometimes he will take advantage of her helplessness by mounting her. Any other males in the harem that are beyond puberty may have a similar response.

Mares have some control over when they deliver their foals. Eighty percent of mares foal at night, usually shortly before dawn, when darkness makes them less noticeable and predators are sated from a night of hunting. If unusual activity or potential danger disturbs the mare, she may delay the birth until she is alone. One study of domestic mares showed that 10% have difficulty foaling, usually due to a malpositioned fetus (. In the wild, when a foal is undeliverable, both mare and foal often die.

When birth is imminent, barrier island mares typically leave the herd and sequester themselves in the marshes or in dense thickets. Sometimes a group of her closest mare-friends will accompany her and appear to offer moral support.

Following the band to the lower lawns, a young colt hesitates at the top of a flight of stairs on the Dungeness grounds. His mother gave him an encouraging nudge to start him on his way, then led him down the incline.

During the first part of labor, the cervix dilates. The mare becomes restless and sweaty, looking at her sides with a concerned look and pacing for about 1–4 hr before delivery.

Delivery typically begins with the breaking of the water sac. During the expulsive process, the mare cannot stop delivery and is consequently vulnerable. The mare lies on her side and pushes repeatedly by contracting her abdominal muscles to assist the expulsive efforts of her uterus. She may rise and then lie down again.

Glistening white membranes appear at the vulvar lips. The foal's front hooves usually present first, one slightly ahead of the other, padded with rubbery keratin to prevent injury to the mare before and during birth. The nose of the foal is positioned atop the forelegs. Further contractions bring the shoulders, body, hips, and hind legs over the course of 15–20 min. With a final strain, the mare deposits the foal on the ground. Mother and newborn rest for a few minutes, breathing heavily.

The mare cranes her neck to glimpse the newborn lying behind her, and bonding begins. She nickers softly to the new arrival, fills her nostrils with his scent and begins to lick him ambitiously. Maternal licking helps to dry the coat, and the vigorous stimulation encourages the newborn to breathe, move, and stand. Licking also acquaints the mare with the odor of her foal and facilitates bonding. Chemicals in the mare's saliva transfer to the fuzzy wet coat and help her to identify her offspring by smell. It is crucially important for the mare to understand that the foal is hers and accept responsibility for his care. If she rejects him, he dies. A mare will not nurse a foal that is not her own.

Some of the foal's behaviors are instinctive, and some are learned. Soon after reaching the outside world, the foal's imperative is to stand. The longer he is immobile, the more likely he is to be devoured by some marauder. He comes into the world with the overwhelming instinct to lurch to his feet, unfolding great lengths of spidery legs and assembling them beneath himself at reasonably appropriate angles.

The umbilical cord breaks when the mother or foal stands after birth, and the placenta is expelled soon after. Most mares will not eat the placenta.

The next objective for the unsteady, spraddle-legged foal is to find sustenance. His instinctive programming prompts him to thrust his muzzle into any dark corner until he finds a teat. He randomly probes the shadows on the underside of his dam, using smell and touch to find his target. The mare stands unmoving to facilitate his search, hind leg pulled back to expose the target, gently guiding him as he noses into her elbow and flank before reaching his goal. At birth, foals sport a bouquet of long, silky whiskers that function as sense organs to help them find an udder. Eventually, he makes contact with one of her two teats and suckles enthusiastically.

Within 2 hours the foal has fed, and by dawn of his first day he can keep up with his mother as she rejoins the herd. At first, he follows his mother with great clumsiness, but this is short-lived. Having mastered

walking, he soon attempts an awkward run, often punctuated by playful hops and bucks. This is quite a feat, considering that hours before he was unacquainted with balance, sequential leg movement, and the inexorable pull of gravity.

Colostrum, thick yellow early milk, is liquid gold. Its immunologic benefits are so profound that the foal is likely to die of infection if he fails to receive it. Immunoglobulins, or antibodies, are proteins of the immune system, produced in the bone marrow of an adult horse and circulated in the bloodstream, specifically in the plasma. There are many kinds, each designed to bind to specific bacteria, viruses, or foreign proteins to target them for destruction. Where human infants acquire some defense against disease from maternal immunoglobulins transferred during pregnancy, foals are born with minimal circulating antibodies in their blood and must suckle immediately after birth to protect them until their immune systems mature. This antibody cocktail is customized for that particular foal in that particular environment. The mare passes antibodies against

Foals, like human infants, sleep often and spend much of their naptime in REM sleep.

diseases that she has been exposed to, which are the diseases that the foal is most likely to encounter.

The mare produces about 5 L of colostrum in the first 12–18 hr. When the newborn nurses, the large antibodies in colostrum are not digested in the foal's small intestine, but are absorbed directly into the bloodstream by specialized epithelial cells. The capacity to absorb antibodies is greatest during the first hours after birth and quickly declines. After the first day, "gut closure" occurs and this mechanism is no longer functional.

Foals are born with the instinct to follow any large moving object. The mare's task is to make sure that she is that object and to keep other horses at a distance for the first few days. For the first 2–3 months, the mare looks after the foal, waits for him to catch up, guides him, and protects him from danger.

When he is a little older, she is less patient with his transgressions, and it is the foal's responsibility to keep up with the herd and find his mother. Likewise, a mare is very tolerant of a newborn's awkward attempts at nursing, but if an older foal is rough or clumsy she will discipline him or walk away.

Newborn foals try a number of nursing positions before settling on the one that yields the best results: side by side with the mare, facing

the rear, in close contact. Newborns nurse about five times an hour for the first five weeks. Foals older than 10 weeks nurse only about once an hour. A hungry foal will sometimes walk alongside his dam, then move directly in front of her, slipping under her neck so she is forced to stop and let him nurse.

As long as a foal continues to nurse, a mare will produce milk. In contrast to domesticated horses, which are usually weaned before 6 months of age, wild horses nurse until they are yearlings if the mare foals again the following year and often until age 2 or even 3 years if they do not have a new sibling. In the wild, horses have been known to nurse for as long as 5 years as a comfort behavior, not for nutritive benefit. Occasionally a mare may allow both the yearling and the new foal to nurse. The author observed one Assateague mare was still occasionally suckling her dam while she had a young foal of her own!

Typically, the other horses in the band are extremely curious about the new arrival. A mare with a new foal is very protective and will temporarily rise in the pecking order as she assertively keeps even dominant horses from her newborn. Maternal protectiveness persists for the first few days and serves to make the foal absolutely sure which mare is his dam while insulating the foal from potentially lethal kicks and bites. The stallion may circle the mare and foal and keep the other horses from approaching too closely.

Young foals engage in coprophagy, or feces-eating. It seems that consuming the manure of adult horses, especially their mothers, establishes the proper balance of helpful intestinal bacteria and helps them to learn which plants to eat.

A newborn foal spends much of his time sleeping or nursing. Unlike mature horses, foals take frequent long naps flat on their sides in deep slumber. When a young foal sleeps, the mare positions herself close to him in a behavior called the "recumbency response." She may graze nearby, mowing the grass short in a circle around her foal, or doze herself while standing beside or over him. Foals will often sleep beside one another in a sort of equine kindergarten naptime, limbs or necks overlapping.

Foals are born with a fleecy coat made of long, soft, curly hairs, which begin to shed by about 7 weeks of age. The molting of the baby fuzz takes about 30–40 days, during which patches of mature coat show in odd places around the eyes and muzzle, often in a different color. The fuzzy upright newborn mane gives way to a more mature growth that stands up comically before it begins to lie down on one side of the neck.

Newborn foals taste grasses with toothless mouths and chew without ingesting, initially not discriminating against poisonous plants.

They do the same with driftwood, sand, and stones. Over the next months, foals learn which foods are palatable and spend more time grazing. The diet of the Cumberland horses centers on five kinds of grass. *Spartina alterniflora*, saltmarsh cordgrass, is the primary item on the menu.

Foals also ingest manure for the first three or four months of life, especially the fresh dung of the mother. This appears to be necessary for establishing optimal intestinal flora and helping foals identify desirable dietary plants.

Two pairs of incisors begin to emerge in 6–8 days, allowing the foal to nibble at grass. Two more incisors emerge by 2–4 months, and by 6–8 months the foal has a full set (six pairs) of deciduous "baby" teeth.

It is possible to determine the age of a horse, especially a younger one, by his teeth. Like children, horses have deciduous teeth followed by permanent teeth. At age 3, the central four incisors are shed; at age 4, the four teeth lateral to these; and at the age of 5 she has all her adult incisors. After the age of 5, other features indicate age, such as the presence and depth of cups on the incisors, dental stars or exposed pulp, and the shape and angle of the tooth. Canine teeth, also known as tusks or tushes, appear in the gap between the incisors and the molars of male horses at the age of 4–5 years, but only occasionally develop in mares.

Foals, like all young mammals, have an innate curiosity and playful spirit. Scientists define play behavior as activities having no immediate use or function and involving a sense of pleasure. Play develops strong bodies, provides practice for adult behaviors like mating and fighting, and builds social and communication skills.

Solitary and object play develop in the first weeks of life while foals remain close to their mothers. By the age of 4 months, foals become more independent. As maternal protectiveness diminishes, foals spend more time in social play, seeking out playmates close to their own age and size. Through their play, they develop coordination and balance, become socialized, and refine behaviors necessary to survival. They gallop madly in circles around the band, not always gracefully, and mimic adult behavior.

Colts play at sexual activities, climbing astride their playmates, mothers, and herdmates with erections but no penetration. Fillies have less interest in sexual play.

Colts with same-age peers develop a dominance hierarchy among themselves based mostly on birth order. Older colts are dominant, as are the sons of high-ranking mares. These buddies play-fight boisterously and spend much more time running than do colts raised singly. This roughhousing gives colts with same-age friends a biological advantage, allowing them to develop greater skills, motor abilities, and physical conditioning than colts raised alone. This aptitude leads to higher rankings in bachelor herds and probably increases the likelihood of acquiring and maintaining a harem.

Fillies romp, too, but do not usually engage in mock battles with the same enthusiasm as colts. They play running games, but not to the extent their brothers do. Because of these differences, fillies prefer to play with fillies, colts with colts.

In males, playful behavior peaks between the ages of 2 and 4. Adult stallions will play with their adult sons and with other stallions outside the breeding season. Young mares are just as playful as colts living in a bachelor band, but lose interest in most play once they have had their first foals.

The herd accepts and indulges male foals for their first two or three years. They can get away with outrageous behavior and social blunders and most members of the herd regard them with overt affection. Colts enjoy playing with the harem stallion and may seek him out for

a sparring match or mutual grooming. They impishly chew the legs and tails of adult horses, gallop disruptively through the band in a gang, break up mutual grooming sessions, and prevent adults from napping. They spend their days romping, wrestling, grazing, nursing, and sleeping. Life is good.

The colt begins to feel the effects of testosterone somewhere around his second or third spring. Mares suddenly seem very interesting. Colts are physically able to impregnate mares by the age of 2–3, but are not behaviorally mature until the age of 5–6. Keiper (1985) writes that 57% of Assateague horses leave their natal band between the 12th and 24th month. Studies of the horses of the Granite Range in Nevada showed that colts dispersed when they were just over 2 years old, with an age range of 11–52 months. This dispersal is usually voluntary; Rubenstein wrote that on Shackleford Banks, mares and stallions rarely drive off juvenile males (Rubenstein, 1982).

Only stallions in good condition rise to the top of the social hierarchy and acquire bands of their own. Because a harem band is dominant and has access to better forage than would a bachelor band, a colt who remains with his natal band until he is 3 or 4 is often larger, stronger, and more confident. This gives him a competitive advantage over colts dispersed at a young age. Adolescent males over the age of 2 often leave and return to the natal band at will until they eventually abandon the band permanently. Some males remain in their natal bands as subordinate stallions and inherit the mares if the harem stallion dies or is displaced.

Sometimes, stallions (or mares) will begin to harass the colt deliberately. The first approaches are not overtly aggressive. The stallion begins to punish the colt for minor infractions, biting at the legs and neck in a less-than-friendly way, like a manager motivating an employee to quit to avoid firing him. After holding an established, secure spot in the social ranking for years, suddenly he is not wanted. Most of the time, the youngster comes to understand his new status and sets out on his own.

At first, displaced colts move to the boundary between the home range of their natal band and that of a neighboring band. Neighboring stallions attack and bully the young trespassers, causing them to move on to new ranges—where the stallions of other bands chase them away.

Horses are gregarious animals and are uncomfortable without others of their own kind. Evicted colts from the same natal band congregate with those from other groups to form bachelor bands, usually of two to four individuals but in some populations as many as 16. Bachelor bands are transitional groups that serve as surrogate family for their adolescent years.

Bachelor bands do more than provide adolescent colts with companionship. Through the frequent mock battles that make up much of their playtime, they learn the techniques necessary to defend a harem, win contests against other stallions, and establish their rank in the dominance hierarchy. They shoulder and shove one another, mutually groom, and stand with their bodies touching, taking comfort from the physical contact. Occasionally they sneak into a harem band and mate with mares in estrus.

Between the ages of four and seven, the bachelor stud reaches full maturity. If he is sufficiently assertive, he begins to steal mares from other stallions to form the nucleus of his own band or challenges harem stallions for herd ownership. Young stallions will often break from a bachelor band and spend some time alone before acquiring a harem.

A stallion (right) harasses an adolescent colt, chasing him away from the watering spring, charging him, biting him, and punishing him when he investigates mares in estrus. The frightened colt responds with submissive gestures. Soon, the colt will leave the band, most likely to join a gang of bachelors in similar circumstances.

The dispersal of adolescents serves to reduce inbreeding. If young horses encounter their sires more than 18 months after leaving the band, they are treated as strangers—colts are regarded as rivals and fillies courted as potential mates. The incest rate varies from herd to herd. A study of Granite Range horses noted that about 4% of copulations were between father and daughter, though these matings are less common in other populations.

Adolescent fillies leave their natal bands of their own accord, usually during a heat cycle when sexually attracted to an unfamiliar stallion. Research on the feral fillies of the Camargue region of France revealed no sign of weakening social bonds with herdmates before leaving. There was also no evidence of sexual competition between the fillies and the mature mares. In fact, the fillies refused sexual advances from the dominant stallion; and if the stallion showed persistent interest, the filly's mother would often intervene and drive him away with back-kicks.

Although some stallions retain daughters in their harems as mates, observations of wild horses in Nevada showed that stallions typically do not court adolescent fillies within their own band, even if the fillies are not their offspring. One study describes several instances of adolescent fillies who copulated with unfamiliar stallions while their fathers stood nearby, unconcerned. More commonly, the

stallion will drive away the filly and her consort. When a stallion is familiar with a pubescent mare, he is not usually interested in mating with her; but if she leaves the band and returns as an adult, the stallion appears not to recognize her and courts her as fervently as he would an unrelated mare.

The dispersing filly will usually join another harem or keep company with a bachelor group. She has a choice about which band to join, inviting herself by wandering close while in heat and encouraging the interest of the harem stallion. Sometimes the stallion drives her away, sometimes he investigates the romantic possibilities. Fillies prefer a dominant stallion that can protect them from the harassment of others who wish to copulate. Often the mares in the new harem do their best to drive a new filly away. If the stallion welcomes the new addition, he will support her when the mares try to exclude her.

Sometimes a mixed group of colts and fillies from the same band will stay together. These adolescent bands are temporary, and their composition changes. Young stallions can acquire mares by claiming a newly dispersed adolescent filly, by challenging and besting a harem stallion, or by staying on the outskirts of a band until they can conquer the stallion or abscond with a number of his mares. As the colts mature, they may depart one by one to form their own bands by stealing mares from harem stallions. Eventually one bachelor is left with the adolescent fillies, allowing him to develop a band with them as his foundation mares.

Typically, young stallions are extremely attentive to newly acquired mares, nuzzling them, grooming them, and following them everywhere. They become the focus of the stallion's whole existence, often more important than even grazing.

Young mares are likely to be happy in their new herds, but older mares often have strong bonds with one another. When stolen from a harem, they may constantly attempt to escape and rejoin their mare-friends. As other mares enter the harem, the stallion will copulate with and defend all his mates, but will typically forge powerful attachments with only one or two mares.

Mature stallions are generally intolerant of unrelated male horses. They do not lay claim to territories, but if a rival dares to invade the personal space of his band, he is obliged to defend it. On the Granite Range of Nevada, 96% of the stallions carried scars from bite wounds. Swiss researcher Claudia Feh (1999) counted battle scars on 20-year-old Camargue stallions—one carried 35, another had 146. Sometimes stallions lose their harems or even die as a result of wounds sustained in altercations with other stallions.

Although stallions are capable of vicious combat, they come to blows only as a last resort. Even fairly superficial wounds can cause lameness, infection, or death, and stallions avoid physical violence whenever possible. Bloodshed is reserved for truly important issues such as the right to mate with a mare in heat. Most of the rivalry between stallions is resolved through ritual displays.

A confrontation between stallions usually begins with the two participants standing well apart, watching each other intently, head raised, ears swiveled forward. Sometimes both stallions will defecate while watching each other at a distance, then move on. This seems to indicate that each accepts that the other owns the mares that accompany him.

"Stud piles" serve as a record of which stallions have recently passed. These layered manure piles are often seen along well-traveled pony trails. Stallions stop and smell them thoroughly, curl their upper lips back in a flehmen response for further olfactory investigation, pivot around or step over the piles, depositing their own manure on top, and sniff again. Flehmen is an odd-looking posture

Horses use the beach for insect relief, but their activities there do not cause significant damage to the natural dune system on the island. This young stallion is a bachelor, unaffiliated with a band.

in which he stretches his neck up to full length, curls back his upper lip, and inhales deeply. This stance allows the inhaled air to reach the vomeronasal (Jacobson's) organs, specialized olfactory devices within the hard palate that can decipher complex olfactory messages. A horse can identify the scents of the other horses that have used the pile and apparently recognize their rank in the social hierarchy. Keiper (1985) writes that stud piles on Assateague can grow more than 6 ft/1.8 m) long and 2 ft/0.6 m) high. If a bachelor herd encounters a stud pile, members will often defecate on the pile in order of rank, from lowest to highest.

If the conflict is not resolved at a distance, the stallions approach each other, posturing to determine which is dominant. After much squealing and striking the air with a fore hoof, one of the two will usually defecate, a sign of subordination. The dominant stallion will then defecate on top of his rival's manure, demonstrating higher rank by allowing his scent to prevail. Each horse sniffs at the manure pile. If the subordinate agrees that the other horse is dominant, they both retreat. Often, this alone somehow resolves the conflict.

If the issue remains unresolved, each stallion attempts to intimidate the other by appearing powerful and assertive. They approach

The home ranges of these Assateague stallions overlap, and they encounter each other regularly. Rather than engaging in battle, they display ritual behaviors that proclaim, "I'm a tough guy, you don't want to mess with me!" Upon meeting, they sniff muzzles, necks, and flanks, their teeth and front legs facing the opponent in case bloodshed is necessary. After about 10–20 seconds of sniffing, both squeal. Another round of sniffing, another pair of ear-splitting squeals. The encounter might go on for 5–10 minutes. The one that yields is the subordinate.

each other with necks arched, tails high, and light, airy steps, then jog shoulder to shoulder in an exaggerated trot known as a parallel prance. In these encounters, they appear to be sizing each other up, and if one is clearly weaker or less confident, the encounter ends as the subordinate stallion withdraws.

If the two stallions still feel equally matched, the contest intensifies. The two rivals sniff at each other, especially at the muzzle and genitals and under the tail. Loud squeals and screams punctuate these investigations. The horse with the loudest, longest squeal often wins the conflict at this point.

If not, they begin to shove each other with their weight and bite at each other, especially at the legs. The fights can quickly turn ferocious. Pulling back, they rear and strike with the forelegs—this is often a display of power rather than a means of inflicting wounds. By rearing and lunging, a stallion can knock the opponent off balance. If the conflict escalates, back-kicking and biting can inflict serious injury.

Some alpha stallions allow a subordinate stallion to remain with the band. In this Assateague band, the pinto is the harem stallion, and he spends considerable energy affirming his dominance over the chestnut. In this picture, they were investigating manure left by other stallions. The chestnut then defecated on the pile, indicating "this is my home range." The pinto defecated on top of the chestnut's dung, signifying his dominance. Then they both turned to thoroughly sniff the pile again.

When the battle is resolved, each stallion rejoins his mares and drives them away from the rival.

Stallion conflicts can last from a few minutes to the better part of an hour, the action taking place in rounds, or battles, separated by retreats to the band to graze. Eventually one of the stallions will give up and take his band elsewhere.

When a stallion grows older and slower, a younger rival usually displaces him. Sometimes the exiled stud will join a group of bachelors, but many old patriarchs become loners. A horse's drive for companionship is powerful, so this imposed isolation must be torture. Stress makes the outcast susceptible to disease and shortens his lifespan.

Experienced stallions can often bluff their way through battles with younger, stronger rivals, retaining control of a herd into old age. Keiper (1985) writes of a 21-year-old stud named Voodoo who kept a harem of seven mares and their foals despite his poor physical condition. Postmortem examination revealed that his bottom incisors had been long since worn away so that the root spaces in his gums had actually healed over.

Courtship among horses is also ritualized. Each of the pair follows a set sequence of behaviors and responses. A mare not fully in heat is unlikely to tolerate a stallion's sexual advances, so he is very careful to test her receptivity thoroughly before attempting to mate. If the stallion is too hasty and crowds a mare, she is likely to kick violently.

Ovulation is seasonal in the mare and influenced largely by length of daylight. In the northern hemisphere, the breeding season is from March through August, peaking in May and June. As spring and summer approach, days lengthen and the mare begins her heat cycles. Free-roaming horses of any given population tend to develop a breeding pattern that begins and ends more abruptly than that of domestic mares. The goal is to foal at a time when the foal is more likely to survive—after the harshness of winter subsides, but early enough in the season that the foal is well developed and fairly independent by the next cold season.

Just before a wild mare comes into estrus, the stallion usually attends her ardently, perhaps keeping her away from the rest of the band or at least other males. He monitors her every move and rushes to smell and scent-mark her manure and urine. He approaches her with a posture that conveys sexual excitement: arched neck, high tail, and flared nostrils. He nickers winsomely. She encourages his attentions by raising her tail invitingly and spreading her hind legs. A mare in heat rhythmically winks her vulva and flashes her clitoris while excreting small amounts of urine. The stallion will touch his muzzle to the puddle and perform flehmen to detect hormones in her urine.

The stallion "teases" her to determine readiness, sniffing, nuzzling, and licking her muzzle and nickering beguilingly. He licks and nibbles her shoulder, elbow, belly, and udder, then hindquarters and genitals. If she is ready for mating, the mare assumes a sawhorse-like stance for mounting with the tail deflected to expose her genitals. The actual mating takes less than a minute. After copulation, the stallion often loses interest and resumes grazing, but the stallion and mare remain close and mate frequently over the next few days.

Mares are receptive to mating for about six of every 21 days during the breeding season, but this varies from mare to mare and with the same mare from season to season. Mares can be fickle, soliciting the

A stallion attempts to intimidate a rival by approaching with an animated prance to display his power and dominance.

stallion's advances one minute and rejecting them the next. A mare will show signs of estrus before she is fully in heat, and if approached at this time, she will reject the stallion's advances with squealing, striking, kicking, and tail lashing. She is most receptive near the time of ovulation and goes out of estrus 24–48 hr after she ovulates.

The sexual instinct is present even among very young colts. Male foals test their mother's urine in the flehmen posture and ritualistically urinate over her puddle. Male foals will attempt to mount other members of the herd, climbing aboard sideways or from the front, bouncing on other males as well as fillies, adult mares, their mothers, or their older sisters. This play-acting eventually evolves into adult sexual behavior.

Horses communicate largely with postures and gestures. This stallion is driving a mare back to his band by lowering his head and snaking it horizontally, ears flattened.

In contrast, fillies do not display sexual behavior until their first heat cycles. Mature Assateague mares go into heat from about March through September if they are not pregnant. They come back into heat about 7–10 days after bearing a foal. If reproduction is left entirely to nature, an average, a mare produces her first foal by age 3, and a stallion fathers foals after the age of 5 although stallions sire most of their progeny after the age of 10.

Dr. Robin Goodloe focused on the Cumberland Horses as part of her doctoral research and became intimately familiar with their habits and behavior (Goodloe et al., 1991). What she saw convinced her that the horses do not belong on Cumberland Island, and based on ecological and genetic considerations, she recommended that the Seashore reduce the size of the herd or remove it entirely. Laliberté similarly concluded "feral horses and hogs are having a disastrous impact on the island's ecosystems, particularly the salt marsh" (2007, p. 44).

Dr. Mark Lenarz conducted the first Cumberland Island free-roaming horse census in 1981, recorded 144, and estimated 10–20 animals were uncounted ("Feral horses," 1983). From 1986 to 1990, a research team headed by Goodloe performed annual horse counts

and found the population steadily rose, Increasing from 144 in 1981 to 186 in 1986 at an annual rate of 5.3%, then dropped to 103 by 1999. Between 2000 and 2011, the Park Service horse census counted 120–154 horses on the island, 148 in 2011. The census for 2014 was 143.

Vincent J. Bellis, aquatic biologist and professor emeritus at East Carolina University, explained the dynamic:

> The number of individuals in a population is thus largely dependent on the available food supply, so that in the absence of human management, such populations can be expected to increase beyond the carrying capacity of the environment. Because migration to better range is seldom an option on a small barrier island, feral populations typically experience cycles in abundance and vigor. A bad year of food production when carrying capacity has already been exceeded can result in drastic losses from to starvation and disease. The survivors may operate with a severely reduced gene pool. Local extirpation may result after several cycles if genetic vigor is not restored by introduction of new individuals. (1995, p. 58)

While breeders of domestic animals keep genetic loss at <2% per generation to avoid inbreeding depression, Goodloe's team, which included renowned equine geneticist E. Gus Cothran, recommended "a more conservative rate of 1%" for most other species, using a

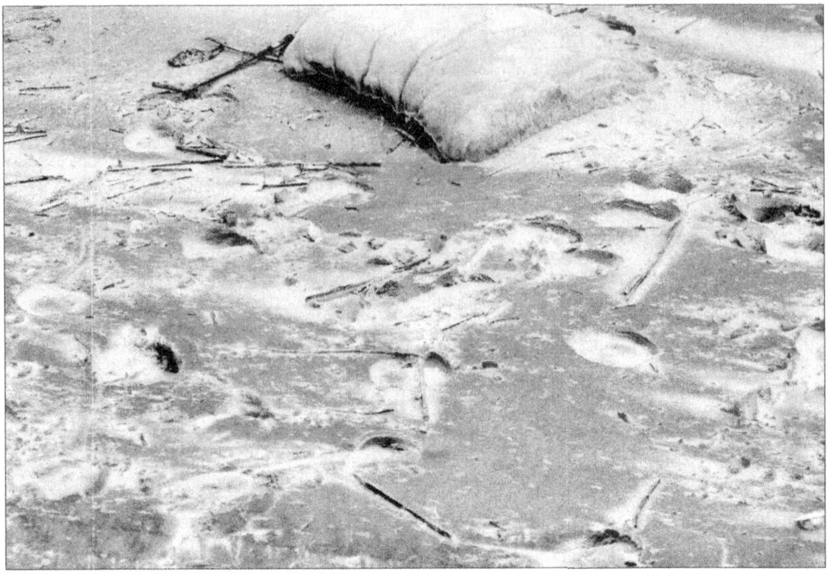

Hoofprints of various shapes and sizes surround a pillow that washed up on the beach. Horses are extremely curious and will investigate, paw, and play with strange objects.

generation length of 10 years for stallions and 9 for mares (Goodloe et al., 1991, p. 419). To prevent genetic loss in excess of this threshold, 23 males and 27 females in the Cumberland Island herd must breed each year. When Goodloe (2000) studied the herd, about 32% of the herd were juveniles not yet reproducing. The team recommended that the current population be halved to 122 individuals, a number that would minimize genetic erosion while reducing pressure on island vegetation. Turner (1988) found that feral horse grazing reduced plant biomass up to 55% and salt marsh vegetation up to 98%. She developed simulation models that indicated an equine population greater than 49–73 would cause serious damage to the salt marshes. A population that size would need outside horses introduced every 9–10 years to minimize inbreeding.

Since 1991, Cumberland Island NS has been monitoring the sex, age, coat color, habitat, body condition, and location of horses and the number of horses in each band through an annual census. Initially, technique differed from year to year. Since 2000, seashore staff has standardized annual counts for the greatest possible accuracy, using the same group of about 30 volunteers during the same month and

The lawns of the Dungeness ruins are favored by the horses as a grazing area, and the home ranges of several bands overlap at this resource. Stallions make concessions to avoid coming to blows. In this picture, the mare in the middle, a mate of the stallion on the right, had strayed to flirt with the harem stallion on the left. The stallion on the left was lower ranking and apparently thought it prudent to return her to her mate. Unwilling to risk the ire of the dominant male, he drove the mare toward the alpha stallion's band. When the alpha claimed her, stepping high with neck arched, the beta stallion defecated, signaling subordinate status. The alpha defecated atop the pile, signifying dominant status. Both stallions are similar in size,shape and markings—they are probably closely related.

same tidal conditions along the same survey routes. The staff then evaluates census data alongside research into horse-related impacts on the island vegetation to develop a long-range plan for horse management. The census does not record every horse—Hoffman estimates as many as 50 horses may be uncounted every year—but the tabulation serves as an "index to abundance" (D. Hoffman, personal communication, 2011).

A population's ability to grow maximally under optimal conditions is its *biotic potential*, represented by the letter *r* in formulas. Usually environmental conditions are not optimal, and increase is limited by environmental resistance—limitations in food, water, climate, habitat, and other conditions. The carrying capacity, the

Most of Cumberland Island is dense maritime forest with a nearly impenetrable understory of saw palmetto and little forage for horses.

number of individuals that a habitat can support indefinitely if food, water, space, and other resources remain constant, is represented by the letter *K*.

For a maximally healthy herd and habitat, managers of wild horse herds should strive to balance a genetically healthy herd size with the carrying capacity of the environment. Carrying capacity (K) is the maximum population threshold that the ecosystem can support, and may vary with climatic fluctuations, i.e., higher with abundant rainfall, lower in a drought. When populations of grazers reach carrying capacity, both herbivores and habitat become stressed and unbalanced. The vegetative environment becomes less diverse, which potentially unbalances the habitat for other species. Oftentimes, dense grasslands are chewed to stubble, trampling becomes widespread, the abundance and distribution of plant species changes, and palatable species are grazed out of existence. When population grows too large, mares begin to produce their first foals later in life, fewer foals and yearlings survive their first winters, and mortality of older animals increases.

A stable horse population, in which births balance deaths and there is no net growth over time, indicates that the herd has reached the

Although Cumberland Island is relatively large, most of it is woodland. Horses forage primarily in the marsh and lawn areas, crowding one another and promoting erosion and overgrazing. Drought reduces the amount of forage available to the horses.

carrying capacity of its range. A herd's proximity to carrying capacity can be assessed by annual census—is it stable, growing, or shrinking?

As the herd reaches carrying capacity, lactating mares, which require more calories than other members, grow uniformly thin; conversely, as the population density decreases, the body condition of lactating mares improves (. The photographs of the lactating mares in this chapter taken during abundant rainfall indicate that when well-watered, Cumberland can support a herd near the top of its historical census without environmental compromise; but when drought conditions prevail, as often they do, there is fierce competition for food resources, and mares become emaciated even if the population is comparatively small.

In studies of domestic stock, a typical horse on pasture has been shown to eat, trample, and damage about 1,000 lb/454 kg of forage per month, and an acre (0.4 ha) may support two average-size horses, depending on season, climate, soil, rainfall, percentage of edible grasses and forbs, and a number of other variables.

Hoffman explained:

> Even if one adjusted this figure to more accurately fit the range of sizes and age classes of the CUIS horse herd, there

Dominance plays a prominent role in nourishment. This mare is a high-ranking member of a high-ranking band. She and her offspring have increased access to higher-quality forage. Consequently, this 2011 photograph shows that she remained in good flesh while nursing her offspring even though she had borne and suckled a colt little more than 2 years earlier. This photograph was taken during a drought of 2 years' duration. Other mares grazing the Dungeness grounds in her vicinity were significantly thinner.

> is still a significant amount of dry matter removed daily that would otherwise be subject to only minimal grazing/browsing pressure by native wildlife. Such pressure exerted over a long period of time can permanently change vegetative communities. (Personal communication, March 18, 2011)

Any species can cause environmental damage through overpopulation. Horses can live on barrier islands without destroying habitat, but they must be maintained below the carrying capacity of the island. If grazing pressures are reduced, balance will eventually return.

In 1995, the Park service assembled twenty-six agency, state, and university specialists to discuss options for managing the herds at both Cumberland Island and Cape Lookout NS. The experts agreed that the horses were causing environmental damage on Cumberland and that the herd should be reduced or removed, a sentiment echoed by several environmentalist groups. One option involved a one-time

Weather has a tremendous influence on the health of the horses and how many the range can support. In 2009, frequent rainfall made the forage lush and abundant, and even lactating mares remained in good flesh.

removal that would reduce the herd to 50–70 animals—the same arbitrary, nonscientific 60-horse population that the government advocated at Shackleford Banks and Corolla. The seashore would use immunocontraception to limit growth and restrict the diminished herd to the part of the island south of Sea Camp, where grass is most abundant and the horses are most observable. A herd of that size, however, is not genetically sustainable, and it would require periodic additions to decrease inbreeding over the long term. Another option was to limit the herd to 120 and allow continued access to the entire island, with immunocontraception to control fertility.

Hoffman explained:

> The same obstacles that make the census challenging would make any of these management options difficult to implement on Cumberland Island. The forests are dense with saw palmetto understory that is difficult to penetrate, complicating roundups and immunocontraceptive administration. Some of the horses living at the south end of the island routinely travel distances of greater than two miles over the course of a day or two. A contraception program would necessitate finding,

In 2011, most of the lactating mares on Cumberland Island were in poor condition. Was their gauntness caused by insufficient food supply, internal parasites, chronic disease, or something else? The root cause was probably overpopulation, which increases the likelihood of the other problems.

> vaccinating, and recording data on every mare on a yearly basis. (Personal communication, March 18, 2011)

In April 1996, the seashore chose the 120-horse option and sought authorization to implement it. At a series of public hearings, a Park Service resource management specialist presented evidence that horses were damaging the marsh and dunes and proposed limiting the population. Her words were a match in a powder keg. Island residents and other locals with strong emotional ties to the herd fervently opposed removing horses from the island. Many speakers questioned the motives of the Park Service, the validity of the studies, and the biases of the researchers. They maintained that the horses had been in residence for centuries and had not caused environmental collapse. They pointed out that wild horses were an integral part of the history, character, and beauty of the island and that their presence increased visitor satisfaction.

Like an emotional magnet, the conflict inexorably drew people in and moved them to one pole or the other. People were passionately for or against the horses. There was no middle ground.

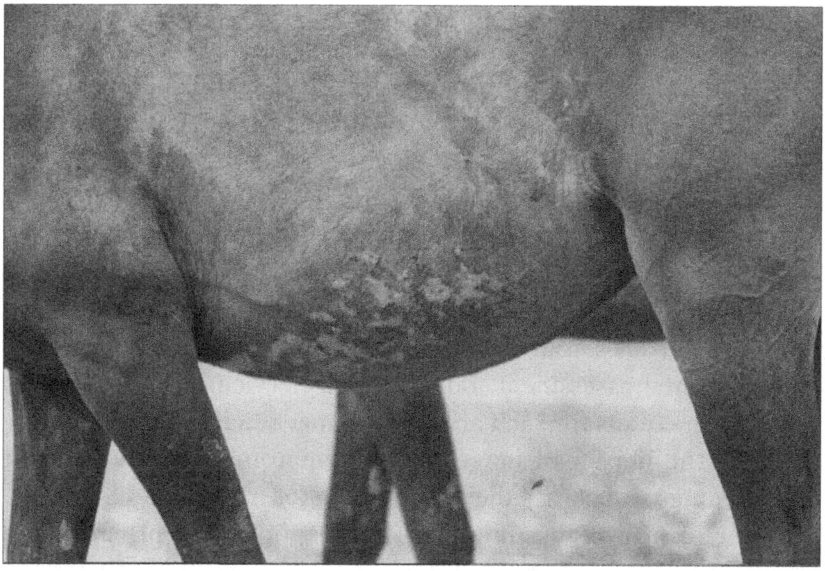

Many Cumberland horses show evidence of what appears to be chronic fungal infection, mostly on the abdomen and across the withers. Widespread fungal infections typically take hold in old, young, debilitated, or immunocompromised animals. They are highly contagious and can spread quickly through a herd.

Opinions fueled agendas, and impassioned arguments broke out in editorials, at home, in the workplace, and of course in the political arena. In fall of 1996, U.S. Representative Jack Kingston (GA 1) toured the island with one of the residents and concluded that the equine census had decreased and there was little evidence of horse-related damage.

Dilsaver recounts (2004, p. 242),

> Without discussing the issue with the Park Service, he added a rider to the fiscal year 1997 budget bill banning all horse management at Cumberland Island National Seashore. The rider suggested that the Park Service use the money allocated for the horse program to further sea turtle protection.

The rider expired the following year, but hampered by budget cuts, the National Park Service has not authored a new management plan. Instead, they have focused on gathering data about the impact of free-roaming horses.

"There are multiple hurdles to creating a management plan for the Cumberland horse herd," says Hoffman:

> We have a very small staff and currently cannot commit the time needed to develop a management plan. Considering the high visibility of this park's horse herd, a concrete and scientifically sound management plan will be needed. This is a very emotional and political issue... there is the possibility that a great deal of effort could go into the development of a new, more current plan, only to be derailed by public pressure and political interests. (Personal communication, March 18, 2011)

A scientifically sound management plan will be based on accurate environmental assessment, but the Park Service says that it does not have the funds for such an undertaking. Scientists are currently studying the impact of horses on the environment in hopes that their findings will clarify the best options for management. A management plan involving immunocontraception is in place for the horses that range on privately-owned Little Cumberland Island, a 2,400-acre tract separated from the main island by Christmas and Brockington creeks.

"If and when the time comes that NPS develops a management plan," said Hoffman, "I envision that we would draw from other management experiences along the Atlantic Coast and likely establish a committee including personnel from various federal and state agencies and private groups that have been involved in the development of plans for the other feral horse herds. Representatives from the local public would be included in the committee as well" (personal communication, March 18, 2011).

The Cumberland Horse Project, a special-interest group organized by Debbie Middleton, a Camden county resident, proposed to collect data as is done on Assateague, Shackleford, and elsewhere. Cumberland Horse Project members volunteered to monitor the condition and social status of each horse; document births, deaths, and social interchange; and assess the health of each horse from a distance. The Park Service cannot afford to undertake this task independently, and these data would lay the groundwork for an evidence-based management plan.

Middleton met with Fred Boyles, superintendent of Cumberland Island NS, and offered the services of her group free of charge. "He stated emphatically that the NPS preference would be to do that themselves," said Middleton (personal communication, January 30, 2011).

Immunocontraception would be difficult to implement on Cumberland Island for the same reason it is difficult to take a census: the horses in the wilderness area tend to avoid human contact and escape into dense woodland when disturbed.

She says that she left that meeting disheartened, feeling that if the Cumberland Horse Project drafted a formal proposal to monitor the health and welfare of the herd for the sake of the horses, it would hit a wall of government pushback.

Nonprofit entities affiliated with the Park Service monitor the welfare of Cumberland Island NS and try to assess its problems. A prominent example is the National Parks Conservation Association, a quasiofficial support group that the first Park Service director and others formed in 1919, just three years after the agency itself came into being. A branch of the NPCA, the Center for Park Research, assesses the strengths and weaknesses of National Parks. The agency's 2009 report (Saxton, 2009), based largely on the research of Jennifer Laliberté and Molli Songco, alleges that equine trampling leaves the marshes highly vulnerable to storm damage and erosion, alters the food web, and negatively affects species such as loggerhead sea turtles.

Yet 2010 was a record year for loggerheads—225 nests a year is the average tally for the island, but in 2010 that number doubled, and hatching success increased. It appears that the horses are not affecting the loggerhead turtles after all. Hoffman offers corroboration:

> Horses do occasionally step on our screens used to protect nests from raccoons, but this is a random occurrence resulting from them walking along dunes and the beach. We have never had a nest predated due to a screen damaged from horse hooves, as we replace or repair the screen soon after it is damaged. (Personal communication, March 18, 2011)

Laliberté's environmental assessment blames the horses for other disruptions as well, stating "The main stressor to the park is the presence of feral horses that graze in the salt marsh, trample the dunes, interrupt sea turtle and shore bird nesting and contribute to erosion" (2007, p. 7). But other wildlife is doing well on Cumberland Island. Recent avian censuses show high counts and excellent species diversity each year (Hoffman, 2008). Bald eagles (*Haliaeetus leucocephalus*) are on the increase. Species uncommon on the island are sighted with increasing frequency, including Franklin's gull (*Leucophaeus pipixcan*), the Nashville warbler (*Vermivora ruficapilla*), and the black-throated blue warbler (*Dendroica caerulescens*). The reddish egret (*Egretta rufescens*), a rare and range-restricted species, has been counted on the island 8 times, all within the last 15 years.

Hoffman, who is intimately familiar with the entire island, sees no indication that horses are damaging dunes. "While the horses do graze sea oats and other plants on the dunes, we have a healthy dune system that is impacted by severe storm, wind, and tide events more than anything else," says Hoffman (personal communication, March 18, 2011).

Horses are implicated in disturbing shorebird nests, but when researchers installed video monitors to determine the cause of American oystercatcher (*Haematopus palliatus*) nest failure, the true culprits were unmasked (Schulte, Brown, & Reynolds, 2007). Raccoons were devouring the eggs, and people walking on the beach near the nests disrupted incubation by frightening birds off the nests. The cameras also caught a young child handling and breaking eggs. "It is likely that horses occasionally step on a nest while walking the beach and dunes," says Hoffman, "But I suspect this is rare" (personal communication, March 18, 2011).

Some reports, including the Cumberland Island NS resource assessment, blame horses for increasing island erosion, which is removing about 2.6 ft/0.8 m of the western shoreline a year. But closer

investigation suggests that the unnatural movement of sand around Cumberland Island is the consequence of two jetties, built to maintain a navigational channel through St. Marys Inlet. The jetties disrupt the longshore and channel flow of sediments, resulting in an unnatural accumulation on the southern tip of Cumberland Island, which has added more than 500 acres/200 ha of new land on the ocean side.

The short-term rate of shoreline change calculated over an approximate 27-year period shows a similar pattern of erosion and accretion—from 396 ft/121 m of accretion in one spot to 346 ft/105 m of shoreline retreat in another. Cumberland Island gained 6.5 ft/2 m a year over a period of about 6 years, a rate consistent with observations over the past 150 years, and Cumberland and Little Cumberland "showed accretion over 80%" of their combined ocean beach. Hoffman says:

> There are certainly some spots where horse activity is impacting shoreline but it is likely not a factor in the major erosion areas. We view erosion as a major issue on Cumberland, with the most significant losses along the western boundary. There are several areas that experience almost constant erosion, primarily as a result of high winds, water currents, and boat wakes. We are planning on conducting an oyster reef

> restoration project in the main problem areas to attempt to slow the losses. (Personal communication, March 18, 2011)

Large sand dunes on the island are migrating and losing height. The back dune ridge at Dungeness dune crossing has decreased in height 13 ft/4 m since the 1960s, and the dune has advanced to engulf trees and spill over into Beach Field at a rate of 3.3–6.6 ft/1–2 m a year. Many blame horse grazing for these changes.

Although overgrazing can destabilize sand dunes, dune migration is a natural process that occurs on barrier islands with or without horses. The movement of the dune at Dungeness dune crossing is the function of sand supply, wind, waves, rain, and the vegetation over which the dune is migrating. The landform is dynamic, and its morphology and rate of migration are ever-changing.

Hoffman offered his perspective:

> Most of our foredunes are well covered in sea oats, suggesting that they can withstand some level of grazing pressure. We have large expanses of dune fields in some places between the foredunes and the forest edge, and there are also large dunes along the forest edge that show signs of migration over a long period of time. I do not think that these dune fields and large inner dunes would have extensive vegetation

growing on them naturally if the horses were not here; I suspect this is indeed a natural process. The sea oats are not clipped off and overgrazed, but appear to be healthy with decent leaf growth and seed production. (Personal communication, March 18, 2011)

As on Assateague, Cumberland horses are blamed for degrading water quality. But other sources point to pollution from nearby watersheds, effects of acid rain, agricultural runoff, and aging septic systems on the island. The levels of mercury in the sound waters are elevated, perhaps by runoff from mercury-based fungicides once used in nearby pine plantations (Alber, Flory, & Payne, 2005). Other possible sources include the defunct Gilman/Durango-Georgia paper mill near St. Marys (a prospective Superfund site) and active paper mills in Fernandina Beach, Fla., and Brunswick, Ga. . Mercury accumulates in the flesh of predatory fish, and there is a regional advisory regarding consumption of local seafood.

Cumberland Island is also threatened by a population explosion on the mainland spurring new housing developments on the coast, together with waterfront mansions and marinas directly across Cumberland Sound from the seashore. In 1980, only 13,371 people lived in Camden County, Ga. By 2000, that number had mushroomed to 43,664.

Global climate change will pose challenges to the Cumberland Island ecosystem. Over the next 100 years, rising sea levels may drown the marsh and push seawater into the aquifers that provide freshwater on the island.

Some of the marshes of Cumberland Island are verdant and healthy, while others are muddy deserts devoid of vegetation. Some studies blame overgrazing for damaging these sensitive wetlands. These mysterious diebacks, however, are not confined to Cumberland Island, and they afflict ungrazed marshes in Louisiana, North and South Carolina, Massachusetts, Connecticut, and Virginia.

As it turns out, the culprit responsible for marsh dieback may be marsh periwinkles (*Littoraria irrorata*), which are most abundant where *Spartina* is dead or dying. The snails feed primarily on fungus. The snails score the stalks of live cordgrass to facilitate fungal infection of the plants, fertilize the wounds with their own dung, and harvest crops of nutritious fungi. Grazing snails and fungal infection kill

A white doe comes to attention atop a dune overlooking a sheltered pocket of maritime forest near the Dungeness dune crossing. In other places on Cumberland Island, dunes migrate like waves of sand, engulfing and killing trees as they progress. Deer and horses are not to blame—dune migration is a natural process and also occurs on islands with undisturbed dunes. Dune movement is primarily the function of sand supply, wind, waves, rain, and the vegetation over which the dune is migrating.

cordgrass and cause marsh die-offs. Horse grazing actually helps protect marshes against die-off by creating conditions favorable to crabs and birds that prey on *Littoraria* snails.

Wildlife ecologist Craig Downer (2011) says,

> For years, it has been the all-too-easy custom for established interests to "scapegoat," or lay the blame, for overgrazing, erosion, threats to native species, and other ecological abuses upon wild horses or burros wherever they occur, by magnifying their impacts out of proportion—all the while conveniently ignoring mankind's own enormously destructive traditions.

Some question the Park Service assertion that there are too many horses on Cumberland. Shackleford Banks, North Carolina, easily supports 110–130 horses though it is much smaller—only 9 mi/14.5 km long and barely 1 mile/1.6 km wide—and not nearly as lushly vegetated as Cumberland. They argue that a population of 150–200 horses should easily maintain a balance with the Cumberland ecosystem.

In the marsh that lies to the south of Dungeness, there is no sign of dieback.

Most of Cumberland, however, is covered by dense forest that contains little horse forage. Hoffman says that although the island is relatively large, the horses can only use certain areas, and large numbers of horses crowd onto them:

> The majority of the island is not suitable habitat for horses, so the areas that have attractive habitat are used by multiple animals. For example, we have approximately 300 total acres of pasture grass-type habitat and most of it is in two places. These see constant grazing pressure. In addition, the salt marsh accounts for approximately 15,000 acres. . . . Not all of the total marsh acreage is accessible to horses due to tidal creeks, deep mud, etc. The areas that are used by horses are used routinely and show grazing pressure and trampling damage. Considering that the island is 36,000 acres, there is really minimal horse-friendly habitat. (Personal communication, March 18, 2011)

The thin lactating mare at the Dungeness dock introduced in the beginning of this chapter suggests that this overcrowding is detrimental to at least some of the horses. In dry years, the majority of lactating mares appear lean and voraciously hungry over most of the island.

Perhaps the least disruptive method for thinning the herd would be to target young horses from well-represented lines for remote

This barbed-wire exclosure in the Abraham Point marsh seems to show horse damage on the outside of the fence and healthy vegetation within. This photograph, however, does not tell the whole story. Along the opposite fenceline, grass is equally tall on both sides. Marsh dieback is evident both within and without the exclosure. The exclosure was installed circa 2001 as part of a graduate research project. When the perimeter was intact, horses somehow jumped into the exclosure and were unable to escape. They died there—the white objects visible in the mud are bones. Researchers cut a section of wire to prevent other animals from getting caught inside, and horses have had free access to the exclosure for years.

sedation, removal, and adoption, as on Shackleford Banks. If horses are removed to the mainland, however, it is likely that another ugly issue may surge to the forefront—the management of equine infectious anemia, or swamp fever.

EIA, discussed at length in Chapter 7, is a retrovirus transmitted by flies from horse to horse. An infected horse has the disease for life and is subject to exacerbations that can make the animal seriously ill and much more infectious than when in the asymptomatic carrier state. According to Georgia law, all horses testing positive for EIA "are to be branded and quarantined for life or disposed of by approved slaughter or euthanasia with burial" (Equine Act of 1969).

When state veterinarians administered the Coggins test for EIA to other free-roaming East Coast herds, the results were devastating.

The herd on Chincoteague National Wildlife Refuge lost 53 of 113 horses; the herd in Rachel Carson Estuarine Reserve (N.C.) lost 9 of 33; Shackleford Banks, N.C., lost 76 of 184, with five more identified and quarantined the following year; and the privately owned herd on Cedar Island, N.C., lost 13 of 15.

The Cumberland herd has never been screened for EIA. It is entirely possible that a number of horses on Cumberland Island carry the disease. If officials made a concentrated effort to test the herd, many animals might be marked for euthanasia. Testing this herd would present a tremendous challenge. On thickly forested Cumberland Island, it is likely that some elusive horses would evade a large-scale gather; and if they carry EIA, they could perpetuate the disease.

The issue could also become socially and politically explosive. The destruction of healthy-looking chronic carriers in great numbers always sparks public outcry and resistance.

Environmentalists hoped that as a national seashore, Cumberland would remain safe from exploitation and would become progressively more primitive. Although the wilderness area designated in 1982 was legally closed to vehicular traffic, most private property owners on the island were permitted to use the main road until their rights ended in 2010. Without vehicular access through the wilderness area, the historic sites at the north end of the island were no longer accessible to visitors unable or unwilling to hike 17 mi/27 km to see them. The Greyfield Inn provided vehicle tours despite the constraints of the Wilderness Act, Park Service regulations, and even a 2004 ruling by the 11th Circuit Court of Appeals (*Wilderness Watch v. Mainella*, 2004) that they are illegal and must cease.

In 2004, island residents lobbied U.S. Rep. Jack Kingston (GA 1) to attach a back-door amendment to the Consolidated Appropriations Act of 2005 (2004), often cited as the Cumberland Island Wilderness Boundary Adjustment Act of 2004. Without any public input, the legislation removed the federal wilderness designation from 25-ft/7.6-m corridors around three dirt roads (Main Road, North Cut Road, and Plum Orchard Spur) and requires the Park Service to maintain them for vehicle access. This was the first time that Congress rolled back a wilderness designation. Conservation

Some areas of marsh are barren mud flats, while others nearby are densely vegetated. Why is the foreground bare and trampled, while thick cordgrass grows in the marsh beyond? Horses have unlimited access to both areas. Although some scientists blame horse grazing for these changes, similar diebacks of *Spartina* have occurred in ungrazed marshes from Cape Cod to Louisiana. New evidence suggests that the culprits are not horses, but snails—namely marsh periwinkles, *Littoraria irrorata* (Bertness, Silliman, & Jefferies, 2004).

and environmental groups were incensed, but unable to overturn the legislation.

The new law required the Park Service to develop a transportation-management plan that would provide *five to eight* bus tours daily to the historic sites on the northern part of Cumberland (Sharp & Miller, 2008). To accommodate the tours, the road required grading, widening, and defoliation. Conservationists argued that the prevailing vision for Cumberland Island NS was for wilderness, not tour buses and vehicular traffic, and that these incursions would detract from the native beauty of the island. The Park Service, however, has a statutory and regulatory duty to make its publicly funded sites accessible to mobility-challenged visitors.

By 2013, the Park Service was running the 30-mi/48-km, 5–6 hour Lands and Legacies tours twice daily, acquainting visitors

with the remains of Robert Stafford's plantation and cemetery, Plum Orchard Mansion, Cumberland Wharf, and The Settlement, remnants of a village established in the late 19th century for black employees of the High Point Hotel. This area, rich in African American history, includes the Alberty House and the First African Baptist Church (1893), which also served as a schoolhouse and community center.

The long-term goal of Cumberland Island NS is "to maintain and restore the natural, functioning ecology of the Island" (Brumbelow, 2007). Natural processes such as wild fires will be allowed to resume, and "alien influences" such as feral hogs and exotic plants will be managed until their influence on the ecosystem is insignificant. The Park Service hopes to reverse the environmental damage done by farming, logging, and other human endeavors while preserving the flavor of Cumberland's unique cultural and historic heritage.

The seashore plans to make improvements to historic sites as funds become available. One project is to restore and maintain the grounds, gardens, viewpoints, roads, buildings and structures of the Dungeness Historic District. Another is to extend the boardwalk over the Dungeness dune crossing to make it wheelchair

The Park Service and a group of young volunteers removed the pews from the First African Baptist Church during a restoration and painting project.

accessible and build a wheelchair-accessible trail into the wilderness area.

Cumberland Island NS has experienced over 40 years of controversy and compromise, pressures from the public, conflict between interest groups, disagreement over management priorities, and contradictions of policy. There is every hope that in time, those who so passionately love the island will find common ground and work collaboratively toward a more harmonious future in which wild horses have a secure place.

References

Alber, M., Flory, J., & Payne, K. (2005). *Water quality conditions near Cumberland Island National Seashore.* Retrieved from http://www.gcrc.uga.edu/PDFs/2005_gwrc_ CUIS.pdf

American Livestock Breeds Conservancy. (2009). *Ossabaw Island hog.* Retrieved from http://albc-usa.org/cpl/Ossabaw.html

Ashley, M.C. (2004). Population genetics of feral horses: Implications of behavioral isolation. *Journal of Mammalogy, 85*(4), 611–617.

Baker, L.A., Warren, R.J., Diefenbach, D.R, James, W.E., & Conroy, M.J. (2001). Prey selection by reintroduced bobcats (*Lynx rufus*) on Cumberland Island, Georgia. *American Midland Naturalist, 145*(1) 80–93. doi: 10.1674/0003-0031(2001)145[0080:PSBRBL]2.0.CO;2

Baron, J. (1982). Effects of feral hogs (*Sus scrofa*) on the vegetation of Horn Island, Mississippi. *American Midland Naturalist, 107*(1), 202–205.

Bellis, V.J. (1995, May). *Ecology of maritime forests of the southern Atlantic coast: A community profile* (Biological Report 30). Washington, DC: U.S. National Biological Service.

Bennett, D., & Hoffman, R.S. (1999, December). *Equus caballus. Mammalian Species, 628,* 1–14.

Berger, J., & Cunningham, C. (1987). Influence of familiarity on frequency of inbreeding in wild horses. *Evolution, 41*(1), 229–231.

Bertness, M., Silliman, B.R., & Jefferies, R. (2004). Salt marshes under siege. *American Scientist, 92*(1), 54–61.

Blum, M. (2013, April 22). Complete forage diet ideal way to feed horses. *AgriNews.* Retrieved from http://agrinews-pubs.com/Content/News/Latest-News/Article/Complete-forage-diet-ideal-way-to-feed-horses/8/6/6832

Boyd, L., & Keiper, R. (2005). Behavioural ecology of feral horses. In D.S. Mills & S.M. McDonnell (Eds.), *The domestic horse: The origins, development and management of its behaviour* (pp. 55–82). Cambridge, United Kingdom: Cambridge University Press.

Brumbelow, J. (2007, August). *The future of America's national parks: First annual centennial strategy for Cumberland Island National Seashore.* Washington, DC: U.S. National Park Service. Retrieved from http://nps01.origin.cdn.level3.net/cuis/parkmgmt/upload/cuis_centennial_strategy.pdf

Bullard, M. (2003). *Cumberland Island: A history.* Athens: University of Georgia Press.

Byrnes, M., & Hiland, M. (1995). Large-scale sediment transport patterns on the continental shelf and influence on shoreline response: St. Andrew Sound, Georgia to Nassau Sound, Florida, USA. *Marine Geology, 126*(1–4), 19–43. doi: 10.1016/0025-3227(95)00064-6

Cameron, E.Z., Linklater, W.L., Stafford, K.J., & Veltman, C.J. (1999). Birth sex ratios relate to mare condition at conception in Kaimanawa horses. *Behavioral Ecology, 10*(5), 472–475.

Carroll, W.D., Kapeluck, P.R., Harper, R.A., & Van Lear, D.H. (2002). Background paper: Historical overview of the Southern forest landscape and associated resources. In D.N. Wear & J.G. Greis (Eds.), *Southern forest resource assessment* (General Technical Report SRS-53) (pp. 583–605). Asheville, NC: U.S. Forest Service, Southern Research Station.

Christmas bird count, Cumberland Island. (2011). Retrieved from http://cbc.audubon.org/cbccurrent/current_table.html

Cofer-Shabica, S. (2010, March). The back dune ridge complex at the Dungeness dune crossing. In O.C. Reedy, F.C. Watts, & F.L. Pirkle, *Quaternary barrier island: Cumberland Island National Seashore.* Guide for the National Society of Consulting Soil Scientists annual meeting scholarship field trip.

Consolidated Appropriations Act, 2005, Pub. L. No. 108-447 (2004).

Crowell-Davis, S.L., & Weeks, J.W. (2005). Maternal behavior and mare-foal interaction. In D.S. Mills & S.M. McDonnell, *The domestic horse: The evolution, development, and management of its behavior* (pp. 126–138). Cambridge, United Kingdom: Cambridge University Press.

Diefenbach, D.R., Hansen, L.A., Warren, R.J., Conroy, M.J., & Nelms, M.G. (2009). Restoration of bobcats to Cumberland Island, Georgia, USA: Lessons learned and evidence for the role of bobcats as keystone predators. In A. Vargas, C. Breitenmoser, & U. Breitenmoser (Eds.), *Iberian lynx ex situ conservation: An interdisciplinary approach* (pp. 423–435). Madrid, Spain: Fundacion Biouniversidad and IUCN Cat Specialist Group.

Dilsaver, L.M. (2004). *Cumberland Island National Seashore: A history of conservation conflict.* Charlottesville: University of Virginia Press.

Downer, C. (2011, March 3). *[Letter to National Wild Horse and Burro Advisory Board][Web log post].* Retrieved from http://tuesdayshorse.wordpress.com/2011/03/06/craig-downer-appeals-to-national-wild-horse-and-burro-advisory-board/

Equine Act of 1969, Ga. Code Ann. § 4-4-110 (2013).

Feh, C. (1999). Alliances and reproductive success in Camargue stallions. *Animal Behaviour, 57*(3), 705–713. doi:10.1006/anbe.1998.1009

Feh, C. (2005). Relationships and communication in socially natural horse herds. In D.S. Mills & S.M. McDonnell (Eds.), *The domestic horse: The origins, development and management of its behaviour* (pp. 83–109). Cambridge, United Kingdom: Cambridge University Press.

Feral horses. (1983). *Park Science, 3*(4), 19–20.

Flueck, W. (2000). Population regulation in large northern herbivores: Evolution, thermodynamics, and large predators. *Zeitschrift für Jagdwissenschaft, 46*(3), 139–166. doi: 10.1007/BF02241353

Gardner, P.E. (Photographer). (1958, April). Front elevation from approach road (northeast)—Dungeness, Cumberland Island, Saint Marys, Camden County, GA. Retrieved from http://lcweb2.loc.gov/pnp/habshaer/ga/ga0200/ga0298/photos/ 055111pu.tif

Georgia Department of Natural Resources, Wildlife Resources Division. (2010, November 1). *Turtle watchers cap 'good year' for loggerheads.* Retrieved from http://www.georgiawildlife.com/node/2391

Georgia toxics release inventory report, 2001. (2003, February). Atlanta: Georgia Dept. of Natural Resources, Environmental Protection Division.

Goodloe, R.B., Warren, R.J., Cothran, E.G., Bratton, S.P., & Trembicki, K.A. (1991). Genetic variation and its management applications in eastern U.S. feral horses. *Journal of Wildlife Management, 55*(3), 412–421.

Goodloe, R.B., Warren, R.J., Osborn, D.A., & Hall, C. (2000). Population characteristics of feral horses on Cumberland Island, Georgia and their management implications. *Journal of Wildlife Management, 64*(1), 114–121. doi: 10.2307/3802980

Goodson, E. (2009, January 14). Is another paper mill auction in the future? Financial defaults lead to higher demands from attorneys. *Tribune & Georgian Online* (St. Marys, GA). Retrieved from

http://www.tribune-georgian.com/articles/2009/01/16/news/top_stories/2topstory1.14.txt

Goodwin, D. (2002). Horse behaviour: Evolution, domestication and feralisation. In N. Waran (Ed.), *The Welfare of Horses* (pp. 1–18). Dordrecht, Netherlands: Kluwer Academic Publishers.

Greger, P.D., & Romney, E.M. (1999). High foal mortality limits growth of a desert feral horse population in Nevada. *Western North American Naturalist, 59*(4), 374–379.

Harlan, W. (2007, August). Line in the sand. *Blue Ridge Outdoors, 29–31*, 34.

Harvey, D., Atteberry, K., Hill, G.T., Fry, J., & Kinzer, M. (2009, April). *Final north end access and transportation management plan and environmental assessment, Cumberland Island National Seashore.* Retrieved from http://www.nps.gov/cuis/parkmgmt/loader.cfm?csModule=security/getfile&PageID=214864

Hoffman, D. (2008, Winter). Mid-winter waterbird survey conducted. *Mullet Wrapper,* p. 4.

Hoffman, D. (2011, March 18). Personal communication.

Hoffman, D. (2011, October–November-a). Current status of the CUIS hog population. *Mullet Wrapper*, p. 4.

Hoffman, D. (2011, October–November-b). Researchers to study CUIS bobcat population. *Mullet Wrapper*, p. 4.

Hoffman, D. (2014, July 20). Personal communication.

Hoffman, D. (2014, July 21). Personal communication.

Johannson, K., & Lindgren, M. (2008). *Infant mortality rate (per 1,000 live birth* [sic]*)*. Retrieved from https://spreadsheets.google.com/spreadsheet/pub?key=0ArfEDsV3bBwCcGhBd2NOQVZ1eWowTnBGMlBUb3YyQ3c&gid=1

Keiper, R. (1985). *The Assateague ponies.* Atglen, PA: Schiffer Publishing.

Keiper, R., & Houpt, K. (1984). Reproduction in feral horses: An eight-year study. *American Journal of Veterinary Research, 45*(5), 991–995.

Kirkpatrick, J. (1994). *Into the wind: Wild horses of North America.* Minocqua, WI: Northword Press.

Kirkpatrick, J.F., & Turner, J.W. (1991). Compensatory reproduction in feral horses. *Journal of Wildlife Management, 5*(4), 649–652.

Laliberté, J. (2007). *National Parks Conservation Association state of the parks report: Natural resource assessment, Cumberland Island National Seashore, Camden County, Georgia* (Unpublished master's project). Nicholas School of the Environment and Earth Sciences, Duke University, Durham, NC.

Larson, M., Rosati, J., & Kraus, N. (2002, February). *Overview of regional coastal sediment processes and controls* (Coastal and Hydraulics Engineering Technical Note ERDC/CHL CHETN-XIV-4). Vicksburg, MS: U.S. Army Corps of Engineers Research and Development Center. Retrieved from http://chl.erdc.usace.army.mil/library/publications/chetn/pdf/chetn-xiv-4.pdf

Lewis, J.M. (2009, February 1). Problems mounting for wild-horse management. *DVM Newsmagazine, 40*(2). Retrieved from http://veterinarynews.dvm360.com/ dvm/Veterinary+Equine/Problems-mounting-for-wild-horse-management/ ArticleStandard/Article/detail/581081

Linklater, W.L., Stafford, K.J., Minot, E.O., & Cameron, E.Z. (2002). Researching feral horse ecology and behaviour: Turning political debate into opportunity. *Wildlife Society Bulletin, 30*(2), 644–650.

Mayer, J.J., Brisbin, I.L., & Brisbin, I.L., Jr. (2008). *Wild pigs in the United States: Their history, comparative morphology, and current status.* Athens: University of Georgia Press.

McCue, P.M., & Ferris, R.A. (2012). Parturition, dystocia and foal survival: A retrospective study of 1047 births. *Equine Veterinary Journal, 44,* Issue Supplement s41, 22–25. doi: 10.1111/j.2042-3306.2011.00476.x

McDonnell, S.M. (2005). Sexual behaviour. In D.S. Mills & S.M. McDonnell (Eds.), *The domestic horse: The origins, development and management of its behaviour* (pp. 110–125). Cambridge, United Kingdom: Cambridge University Press.

McGreevy, P. (2004). *Equine behavior: A guide for veterinarians and equine scientists.* London, United Kingdom: W.B. Saunders.

Monard, A.M., Duncan, P., & Boy, V. (1996). The proximate mechanisms of natal dispersal in female horses. *Behaviour, 133*(13/14), 1095–1124.

Middleton, D. (2011, January 30). Personal communication.

Milanich, J.T. (1996). *The Timucua.* Oxford, United Kingdom, and Cambridge, MA: Blackwell Publishers.

Monard, A.-M., Duncan, P., Fritz, H., & Feh, C. (1997). Variations in the birth sex ratio and neonatal mortality in a natural herd of horses. *Behavioral Ecology and Sociobiology, 41*(4), 243–249. doi: 10.1007/s002650050385

National Historic Preservation Act, 16 U.S.C. § 470 (1966).

Ogburn, M.B., & Alber, M. (2006). An investigation of salt marsh dieback in Georgia using field transplants. *Estuaries and Coasts, 29*(1), 54–62.

Pareja, F. (1627).*Catecismo en lengua Timuquana y Castellana* [Catechism in the Timucuan and Castilian languages]. Mexico, New Spain: Juan Ruiz. Retrieved from http://memory.loc.gov/master/ipo/qcdata/qcdata5/early_americas/tifs/ea0104_02.tif

Rubenstein, D. (1982). Reproductive value and behavioral strategies: Coming of age in monkeys and horses. In P.P.G. Bateson & P.H. Klopfer (Eds.), *Perspectives in Ethology* (Vol. 5, Ontogeny, pp. 469–487). Princeton, NJ: Princeton University Press.

Rubenstein, D.I., & Hack, M. (1992). Horse signals: The sounds and scents of fury. *Evolutionary Ecology,* 6, 254–260.

Saxton, D. (2009). *State of the parks: Cumberland Island National Seashore, a resource assessment.* Washington, DC: National Parks Conservation Association.

Seabrook, C. (2002). *Cumberland Island: Strong women, wild horses.* Winston-Salem, NC: John F. Blair.

Schulte, S., Brown, S., Reynolds, D., & the American Oystercatcher Working Group. (2007, June). *A conservation action plan for the American oystercatcher (*Haematopus palliatus*) for the Atlantic and Gulf coasts of the United States* (Version 2.0). Retrieved from http://www.fws.gov/migratorybirds/CurrentBirdIssues/Management/FocalSpecies/Plans/AMOY.pdf

Sharp, R.L., & Miller, C.A. (2008). Wilderness, wildness, and visitor access to Cumberland Island National Seashore, Georgia. In D.B. Klenosky & C.L. Fisher (Eds.), *Proceedings of the 2008 Northeastern Recreation Research Symposium* (Gen. Tech. Rep. NRS-P-42) (pp. 216–222). Newtown Square, PA: U.S. Department of Agriculture, Forest Service, Northern Research Station.

Stockdon, H., Thompson, D., & Fauver, L. (2007). *Vulnerability of National Park Service beaches to inundation during a direct hurricane landfall: Cumberland Island National Seashore* (U.S. Geological Survey, Open-File Report 2007-1387). Retrieved from http://pubs.usgs.gov/of/2007/1387/ofr2007-1387-CUIS.pdf

Thorn, V. (1977). *Cumberland Island: A place apart.* Washington, DC: U.S. National Park Service, Division of Publications.

Trussell, J. (2011). Contraceptive efficacy. In Hatcher, R.A., Trussell, J., Nelson, A.L., Cates, W., Kowal, D., & Policar, M. (Eds.), *Contraceptive technology* (20th ed.) (pp. 19–47). New York, NY: Ardent Media.

Turner, M. (1988). Simulation and management implications of feral horse grazing on Cumberland Island, Georgia. *Journal of Range Management, 41*(5), 441–447.

U.S. National Park Service. (1955). *A report on our vanishing shoreline: The shoreline, the survey, the areas.* Washington, DC: author.

van Dierendonck, M., & Goodwin, D. (2005). Social contact in horses: Implications for human-horse interactions. In F. de Jonge & R. van den Bos (Eds.), *The human-animal relationship: Forever and a day* (pp. 65–82). Assen, Netherlands: Royal Van Gorcum BV.

Warren, L.K. (2004). *Determining the grazing capacity of your horse pasture.* Retrieved from http://www.coopext.colostate.edu/boulder/sam/pdf/PastureGrazingCapacity.pdf

Wilderness Act, 16 U.S.C. §§ 1131–1136 (1964).

Wilderness Watch v. Mainella, 375 F.3d 1085 (11th Cir., 2004).

Worth, J.E. (1998). *Timucuan chiefdoms of Spanish Florida: Vol. I. Assimilation.* Gainesville: University Press of Florida.

Wright, H., & Lawrence, R. (2012). *Feral animals on Cumberland Island.* Retrieved from http://wildcumberland.org/?page_id=670

Acknowledgments

About 20 years ago, I began to research my first book, *Hoofprints in the Sand: Wild Horses of the Atlantic Coast.* The experience enriched my life in many ways. I learned a tremendous amount about how the natural behavior of these horses differs from that of their domestic counterparts. Following herds through marsh and dune, I gained a greater appreciation of the complexity of their apparently simple lives in the wild. Similarly, with the writing of *Wild Horse Dilemma*, many helpful and extraordinary people have earned my gratitude, and many have become friends.

Dr. Jay F. Kirkpatrick, senior scientist at the Science and Conservation Center at ZooMontana in Billings and author of *Into the Wind: Wild Horses of North America*, has devoted much of his career to preserving the health, rights, and dignity of wild horses. He developed and implemented the immunocontraceptive program in use with many species of wildlife, including horses. He answered my questions, forwarded useful documents, crafted an insightful preface, and generously reviewed the manuscript before publication.

Don Höglund, DVM, is an internationally esteemed leader in horse training and management and the author of *Nobody's Horses*, a riveting book about the rescue of wild horses from the White Sands Missile Range. He has implemented numerous large-scale equine programs, including the Department of the Interior's Wild Horse Prison Inmate Training Program, which teaches prisoners to gentle horses while providing training for adoptable mustangs. His love and admiration for horses is evident in all that he does. When I approached him with a few questions, he responded enthusiastically and sent me a number of articles that shaped the backbone of my manuscript. I am grateful for his support and encouragement along the way and fortunate that he agreed to review the manuscript.

Dr. E. Gus Cothran, Texas A&M University's renowned expert on the genetics of wild and domestic horses, helped me to understand the significance of the Q-ac variant in certain wild horse herds and the concept of minimum viable population. He also found time in his busy schedule to review the manuscript. His research is the cornerstone of wild horse management, and I have cited it extensively.

Dr. Sue Stuska, the wildlife biologist at Cape Lookout National Seashore who oversees the Shackleford Banks herd, has corresponded regularly about the status of the horses. When I visited, she taught me how to identify individuals, and showed me her dynamic census chart that tracks the members of each band and where they were last sighted. She explained how the current management plan makes optimal use of the existing gene pool by monitoring family lineage and contracepting certain mares. Sue also gave of her valuable time to review my manuscript for accuracy.

Doug Hoffman, wildlife biologist at Cumberland Island National Seashore, helped me to understand the Park Service perspective on the horses living there and corrected my assumptions and misinformation. He generously drove me to key parts of the island that I could not otherwise reach, and my time with him was the highlight of the trip.

Karen McCalpin, director of the nonprofit Corolla Wild Horse Fund, Inc., found time in her impossibly busy schedule to meet me and discuss the genetic crisis facing the herd. She also reviewed the manuscript. Karen and the other members of the organization—mostly volunteers—have upended their lives to secure protection for these horses. Karen produced a beautifully written blog highlighting the triumphs and tragedies of the herd. It can be accessed at www.corollawildhorses.com

Dr. Ronald Keiper, Distinguished Professor of Biology (emeritus) at Penn State University, was one of the first scientists to study the behavior of horses in the wild, answered my questions about the foaling rate of lactating mares and shared his groundbreaking research detailing the behavior of the Assateague horses.

Wesley Stallings, former manager of the Corolla herd, took me in his truck several times as he patrolled the Outer Banks north of Corolla, following the movements of wild bands and logging herd data in his notebook. Sometimes we climbed on the roof of the truck to scout for horses. Sometimes Wes climbed a tree for a better view. At one point we encountered a flooded hollow and were forced to don hip boots and slog through surging currents occupied by cottonmouth snakes to evaluate the health of

a newborn colt. I am grateful he allowed me to participate in his daily adventures.

Steve Edwards, by day an attorney for Isle of Wight County, Virginia, works magic in rehabilitating injured Corolla and Shackleford horses. At his farm, Mill Swamp Indian Horses in Smithfield, Va., he teaches children how to train wild horses with natural horsemanship techniques. He also established an off-site breeding program to preserve the herd's rare genes in case of disaster in the wild. Steve has been extremely supportive and helpful throughout the writing of this book, and has brought his expertise to the task of reviewing this book for accuracy before publication.

In 2012, I visited Mill Swamp and was captivated by the sight of children working in the round pen with young horses, many of them recently brought in from the wild. To this point, I had great esteem for the wild horses living on North Carolina's Outer Banks, but had never ridden one. I found these Colonial Spanish Horses astonishingly surefooted, brave, rugged, and smooth-gaited. The climax of my visit was a ride through the inky forest astride Manteo, a wild-born black stallion. He never missed a step despite exposed roots, steep embankments, deep pools, and deer crashing gracelessly through the underbrush. For the better part of an hour, we trotted and cantered through darkness so complete, I could not see my hand in front of my face. I had recently been injured in a riding accident, and I was working through many horse-related fears. It was a profound and humbling experience to trust a once-wild stallion to find his way through darkness that left me blind.

D. Phillip Sponenberg, DVM, PhD, helped me to understand the genetic underpinnings of coat color and its implications for the free-roaming Banker horses. In his review of the manuscript, he offered great insights on Spanish horse origins and genetics, and his comprehensive articles on that topic were a valuable resource.

Carolyn Mason, president of the Foundation for Shackleford Horses, Inc., accompanied me to Shackleford Banks and generously shared her wealth of knowledge about the horses. She introduced me to the Banker Horses grazing in her yard, gentled animals awaiting adoption. My heart melted when a young gelding named Adagio followed me like a puppy and courted hugs and scratching.

Woody and Nena Hancock loaded me and my cameras into their boat and searched island and marsh for members of the Cedar Island herd. They introduced me to Bucky, the most genetically valuable horse on Cedar Island, and her 2-week-old look-alike filly, Gay; a mare who prefers the company of three burly wild bulls; and Shack, the robust sorrel patriarch whose photograph graces the front cover of *Wild Horse Dilemma*. It was a profound, almost holy experience to stand calf-deep in warm estuarine waters under a moody sky, surrounded by peaceful wild horses, splashing pelicans, and wind-licked marsh grass.

Laura Michaels, the Park Service ranger in charge of pony care, took me behind the scenes to meet the Ocracoke horses. I also met Wenzel, Doran, Sacajawea, and Jitterbug, the Shackleford horses who will revitalize the Ocracoke herd. I even scratched the neck of the lovely black-and-white mare Easter Lady after admiring her from afar for years.

Roe Terry, former public relations specialist of the Chincoteague Volunteer Fire Company, invited me to the workshop where he carves graceful wooden waterfowl and discussed the challenges faced by the hardworking firefighters. Besides managing the herd of free-roaming ponies, these dedicated people donate their time to provide tax-free fire suppression, search and rescue, and emergency medical services

in a town of 4,400 permanent residents that receives roughly 1.5 million visitors a year. He also granted me access to the optimal vantage point for the world-famous Chincoteague Pony Swim: whereas most onlookers stood in a field behind an orange fence, out of harm's way, I was able to stand directly on the grassy landing where the horses regained solid ground after swimming the channel from Chincoteague National Wildlife Refuge. Ponies rose out of the water like mythical creatures of the sea, dripping wet and looking very pleased with themselves. My feet were in their hoofprints, and occasionally I dove for cover as a stallion thundered by in pursuit of a rival. It was a magical experience.

Denise Bowden, his successor at the fire company, cheerfully supplied me with useful information. Her passion for the horses and the refuge are evident, and her enthusiasm enhances the overall festivity of Pony Penning week.

Lou Hinds, former manager of the Chincoteague NWR, took me around the refuge to show me unequivocal signs of dramatic environmental change, such as tree trunks, light poles, and chunks of peat that had once been on the bay side of Assateague until island migration situated them squarely on the beach. Studying the dynamic nature of barrier islands is one thing; seeing the evidence of their migration is another thing entirely.

Pam Emge, co-author of *Chincoteague Ponies: Untold Tails,* can identify all of the Chincoteague wild ponies and knows the intimate details of their relationships and lineages. She reviewed part of the manuscript, corrected errors, and filled in details.

Anthropologist Karen Dalke of the University of Wisconsin-Green Bay shared her doctoral dissertation and other writings, which provide unique perspective on what we feel about wild horses and how we define them. She also reviewed the manuscript prior to publication.

Paula Gillikin, manager of the Rachel Carson North Carolina National Estuarine Research Reserve, assisted me in researching the horses of Carrot Island and vicinity.

Philip Howard, nephew of Marvin Howard (1897–1969), who led Ocracoke's mounted Boy Scout troop in the late 1950s, and grandson of the legendary horseman Homer Howard, allowed me to use excerpts from his Web site detailing his family history with the wild ponies.

Allison Turner, biological science technician at Assateague Island National Seashore, supplied excellent information about the Maryland herd and shared Park Service photographs that vividly show the bites and kicks that occur when people get too close to wild horses.

DeAnna Locke, administrator of the Ocracoke Preservation Society, let me pore over and digitize the organization's fascinating scrapbooks, which included many pictures of the island's mounted Boy Scout troop.

Tim Ferry and Flickr user rich701 allowed me to reprint some of their unique historical photographs of the Chincoteague herd.

Craig Downer, a wild-horse ecologist and activist on the board of directors of the Cloud Foundation, shared several of his writings with me on the subject of mustang management.

Jean Bonde of the Buy-Back Babes has a contagious enthusiasm for the Chincoteague Ponies and many tales to tell. These ponies have a large cult following, and her e-mail group recounted the details of their lives—celebrating the romance of Copper Moose and Scotty's ET, pondering Rip Tide's status within Surfer Dude's band, and speculating on why Queenie and Suzy Sweetheart were wandering around the wildlife loop.

Tabetha Fenton of Barefoot Minis helped with proofreading and offered enthusiastic support.

Special thanks to my mother, Joyce Urquhart, who is an exceptionally good proofreader. She read every word of the manuscript and discovered errors that other readers missed.

My husband, Alex, is committed to giving me space in which to create and assisting me wherever possible in my creative pursuits. Besides proof-reading my manuscripts, he is the behind-the-scenes man who maintains the household, runs to the post office, and brings in the bird feeders at night so the bears don't destroy them. He is the love of my life and I am thankful every day that we are together.

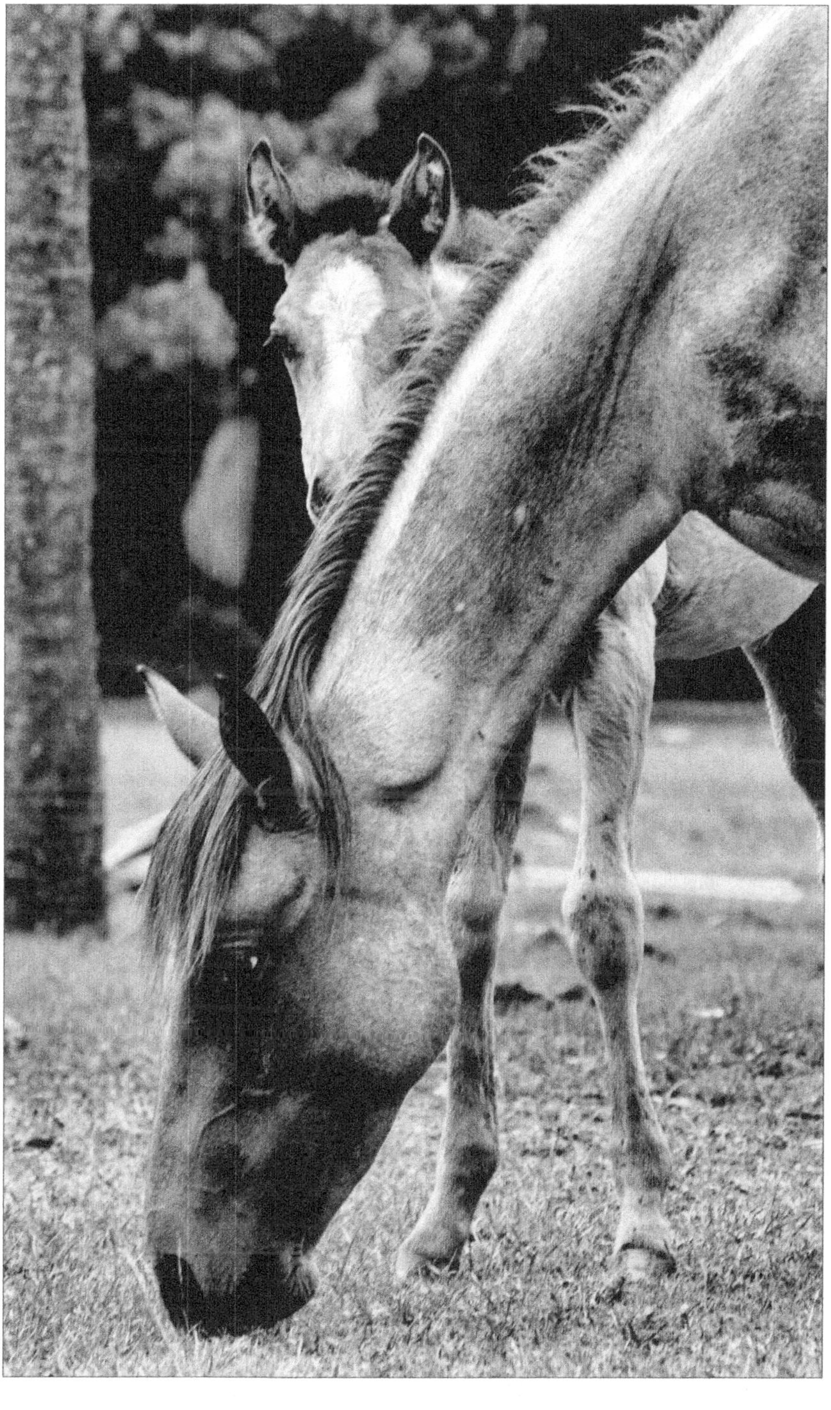

About the Author

Bonnie Urquhart Gruenberg is a multifaceted person who wishes that sleep were optional. She is the author of the award-winning textbook *Birth Emergency Skills Training* (Birth Guru/Birth Muse, 2008); *Essentials of Prehospital Maternity Care* (Prentice Hall, 2005); and *Hoofprints in the Sand: Wild Horses of the Atlantic Coast* (as Bonnie S. Urquhart; Eclipse, 2002), as well as articles in publications as dissimilar as *Equus* and the *American Journal of Nursing*. She is an artist and photographer and has illustrated all her own books.

By profession, she is a Certified Nurse-Midwife and Women's Health Nurse Practitioner who welcomes babies into the world at a freestanding birth center in Lancaster County, Pa. She obtained her MSN from the University of Pennsylvania after completing her BSN at Southern Vermont College, and she spent 10 years attending births in tertiary-care hospitals before returning to out-of-hospital practice. Prior to her career in obstetrics, she worked as an urban paramedic in Connecticut.

Horses have been her passion from infancy. For nearly two decades, she has spent countless hours researching and photographing the private lives of wild horses in both Western and Eastern habitats. She has been riding, training, teaching, and learning since her early teens, from rehabilitating hard-luck horses to wrangling trail rides in Vermont and Connecticut. In her vanishing spare time, she explores the hills and hollows of Lancaster County astride her horses Andante and Sonata.

More information and a collection of her photographs can be found at her Web site, www.BonnieGruenberg.com Additional information about the Atlantic Coast horse herds is on the Web at www.WildHorseIslands.com

If you liked this book, you may enjoy other titles by the author:

The Wild Horse Dilemma: Conflicts and Controversies of the Atlantic Coast Herds (Quagga Press, 2015)
The Hoofprints Guide Series (Quagga Press, 2015)
- Assateague
- Chincoteague
- Corolla
- Ocracoke
- Shackleford Banks
- Cumberland Island

Forthcoming

Wild Horse Vacations: Your Guide to the Atlantic Wild Horse Trail with Local Attractions and Amenities (Quagga Press, 2015)
Wild Horses! A Kids' Guide to the East Coast Herds (Quagga Press, 2015)

Visit QuaggaPress.com for details.

Made in United States
Orlando, FL
10 November 2026

90498885R00066